Charles Frederick Holder

Marvels of Animal Life

Charles Frederick Holder

Marvels of Animal Life

ISBN/EAN: 9783337095192

Printed in Europe, USA, Canada, Australia, Japan

Cover: Foto ©berggeist007 / pixelio.de

More available books at **www.hansebooks.com**

MARVELS

OF

ANIMAL LIFE

BY

CHARLES FREDERICK HOLDER

FELLOW OF THE NEW YORK ACADEMY OF SCIENCES, AUTHOR OF "ELEMENTS
OF ZOÖLOGY," ETC.

LONDON

SAMPSON LOW, MARSTON, SEARLE, AND RIVINGTON

CROWN BUILDINGS, 188 FLEET STREET

TO MY WIFE, THE COMPANION OF MANY OF MY RAMBLES,

THIS BOOK IS AFFECTIONATELY DEDICATED.

PREFACE.

Many of the observations chronicled in the following pages were made during a long residence upon a coral reef or atoll, some while swimming under water along the bristling coral banks that formed a characteristic feature of our tropical home, and others are the memories of many practical collecting tours in various localities in search of specimens valued by the lover of nature.

The chapters relate to widely different subjects; the strange forms of animal life found in various parts of the world, that from the striking evidence of design in their structure and ways are deemed the marvelous productions of the Great Maker.

Natural objects are not appreciated by the majority of young people until the attractive side has been pointed out, and I have endeavored to accomplish this by presenting some of the thoroughly interesting features of nature that will incite my young friends to take to the woods and streams, and become investigators themselves, selecting some of the subjects—mimicry, nest-building fishes, etc.—as their specialties. The way to study the sword-fish is to first go sword-fishing, if you can, and so with all natural studies, take to the field and arouse an interest for the work to be done. I have told about sharks and other animals while giving my own experience with them. in this way hoping to show how much profit and pleasure may be derived from personal observation in these and other subjects.

An attempt has been made to do justice to the fishes, their habits and ways, and to other unfamiliar animals that perhaps have been too long neglected in popular works, and whose claim to our interest is equally strong.

To render many of the topics more valuable and exhaustive, my own observations are frequently supplemented by the opinions of specialists and others, so that my young readers may have access to the interesting and not generally available material relating to the subject.

While I have endeavored to present the facts in simple language, and in a popular manner, the work being in no sense a scientific one, some technical names will be found that are indispensable to any one wishing to determine an animal, as often the common name changes with the locality or is absent altogether.

The subject matter of a few of the chapters originally appeared in Harper's and Lippincott's Magazines, but has been revised and adapted for a wider field of readers, young and old.

C. F. H.

New York, October 1st, 1885.

CONTENTS.

PAGE

CHAPTER I.

THE NEST-BUILDERS OF THE SEA, ...   1

CHAPTER II.

LIVING HOMES, ...   22

CHAPTER III.

DRY LAND FISHES, ...   33

CHAPTER IV.

METEORS OF THE SEA, ...   40

CHAPTER V.

PARENTAL CARE AMONG ANIMALS, ...   50

CHAPTER VI.

AN OCEAN SWORDSMAN, ...   61

CHAPTER VII.

FINNY LIGHT-BEARERS, ...   75

CHAPTER VIII.

OLD FRIENDS, ...   83

CHAPTER IX.

IS THERE A SEA-SERPENT?   91

CHAPTER X.

ANIMAL ELECTRICIANS, ... 109

vii

PAGE

CHAPTER XI.

OUR COMMON SNAKES, ... 120

CHAPTER XII.

ANIMAL MIMICS, ... 138

CHAPTER XIII.

LOST RACES, ... 151

CHAPTER XIV.

THE INK-BEARERS, ... 163

CHAPTER XV.

THE KING OF CRABS, ... 170

CHAPTER XVI.

THE TIGER OF THE SEA, ... 173

CHAPTER XVII.

LIVING LIGHTS, ... 192

CHAPTER XVIII.

WONDERS OF THE AIR, ... 198

CHAPTER XIX.

ANIMAL TRAPS AND TRAPPERS, ... 204

CHAPTER XX.

THE WHITE WHALERS, ... 215

LIST OF ILLUSTRATIONS.

PAGE.

Frontispiece. THE SHELL TRIDACNA AS A PLAY-HOUSE, . . . Face Title

PLATE   I. LAMPREY EELS BUILDING A NEST, ...   1

"   II. GOURAMI AND NEST, ...   8

"   III. PARADISE FISH BUILDING ITS BUBBLE NEST, ... 16

"   IV. HERMIT-CRAB AND ITS BOARDER AS IT APPEARS UNDER

WATER, ... 24

"   V. PERIOPHTHALMI SEARCHING AFTER FOOD ON DRY LAND, 32

"   VI. AMERICAN GOBIES CRAWLING ON THE SHORE, ... 48

"   VII. CLIMBING PERCH MIGRATING OVER LAND, ... 56

"   VIII. SOUTH AMERICAN CAT-FISH CRAWLING ON DRY LAND, 64

"   IX. DRY BURROW OF THE LUNG FISH, ... 72

"   X. MARTINIQUE TREE-TOAD, WITH YOUNG CLINGING TO ITS

BACK, ... 80

"   XI. THE SAILOR-FISH OF CEYLONESE WATERS, ... 88

"   XII. SWORD-FISH ATTACKING A DORY, ... 96

"   XIII. THE BRILLIANT LAMP FISH, ... 104

"   XIV. THE LUMINOUS ARGYROPELETUS, ... 112

"   XV. HERMIT-CRAB IN A TOBACCO-PIPE, ... 120

"   XVI. THE PEMAQUID SEA-SERPENT, ... 128

                                                    PAGE.

PLATE   XVII. ELASMOSAURUS PLATYURUS, FIFTY FEET IN LENGTH, 136

  "   XVIII. AN EXTINCT REPTILE. CAMARASAURUS, SEVENTY-
               FIVE FEET LONG, ... 144

  "   XIX. YOUNG RATTLESNAKES ENTERING THEIR MOTHER'S
               MOUTH, ... 152

  "   XX. INSECT THAT MIMICS A MOSS-COVERED TWIG, . . . 160

  "   XXI. THE WALKING LEAF (PHYLLIUM SICCI FOLIUM), . 168

  "   XXII. A CATERPILLAR THAT MIMICS A SHREW, ... 176

  "   XXIII. MAMMOTH ADRIFT ON AN ICE FIELD, ... 188

  "   XXIV. EXTINCT SEA-COW (RHYTINA), THIRTY FEET IN
               LENGTH, ... 192

  "   XXV. GIANT SQUID, FIFTY-FIVE FEET LONG, ... 198

  "   XXVI. GIANT OCTOPUS, TWENTY-EIGHT FEET ACROSS, . . 204

  "   XXVII. GIGANTIC EXTINCT PTERYGOTUS, COMPARED WITH A
               LOBSTER, ... 210

  " XXVIII. SPOTTED SHARK (RHINODON), SEVENTY FEET LONG,
               RISING UNDER A CANOE, ... 214

  "   XXIX. A GIGANTIC PYROSOMA, ... 216

  "   XXX. GROUP OF FLYING TREE-TOADS, ... 220

  "   XXXI. PTERANODON, AN EXTINCT FLYING REPTILE, . . . 224

PLATE I.

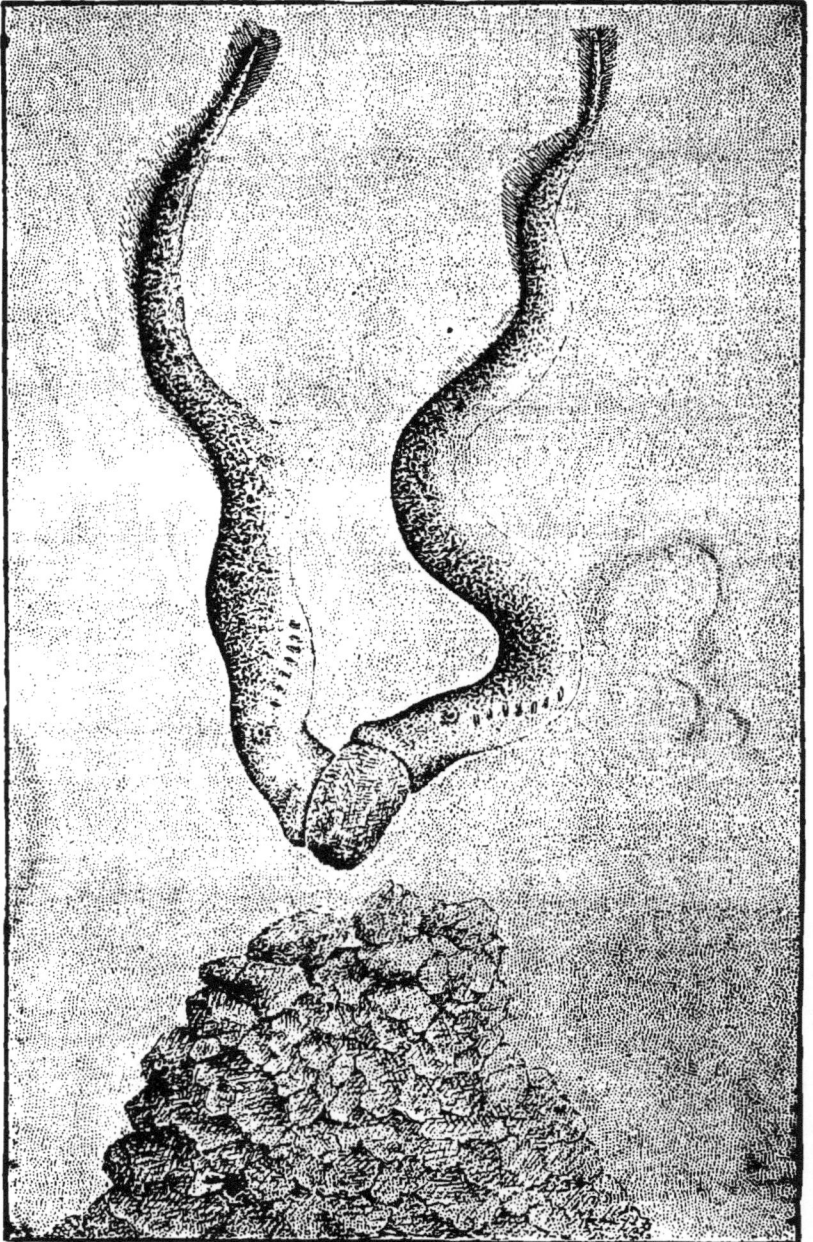

LAMPREY EELS BUILDING A NEST.

MARVELS OF ANIMAL LIFE.

CHAPTER I.

THE NEST-BUILDERS OF THE SEA.

How often we are impressed by the resemblance in the actions and habits of animals in totally different phases of life! In drifting over the reefs of our Southern border, the similarity between the creatures of land and sea is extremely striking. The gardens of the lower world abound in lavish growth: trees, shrubs, waving vines, are all produced in the wondrous forms of the sea. Here a forest of coral branches raise their myriads of bristling points, each tipped by a delicate polyp, and presenting a rich olive-green tint in contrast to the deep blue of the channel, upon whose banks they grow. Pure as crystal, the water seems to intensify the beauty of the objects, even in the greater depths; gaily-bedecked fishes move lazily about; rising and falling among the living branches; poising, perhaps, to pluck some morsel from a limb, in all their motions reminding us of the birds of the shore. The gorgeous parrot-fishes are the sun-birds of the sea; wondrous tints—azure-blue, golden-yellow, and red—mark them. Some appear iridescent, bathed in metallic hues, as if encased in burnished armor; while many more in modest garb, found in our colder waters of the North, call to mind the robin and the thrush, those welcome harbingers of spring. But it is not in their color alone that the fishes re-

1

semble the birds ; it is in the home-life and love of offspring
that we find a close resemblance. Many are nest-builders, erecting structures as complicated as those of some birds, and equalling
them in design and elaboration. In floating along the shores of
some woodland stream, we may watch the domestic life of the
sun-fish (*Eupomotis*), the mottled, bespangled plague of the
angler, that is seemingly always on hand to be caught in default
of nobler game.

Along the borders where delicate grasses grow, where floating
lily-pads cast their shadows, and the white-pink-tipped buds
reach upward, here among the winding stems, perhaps sheltered
by a mossy bank with overhanging ferns, the sun-fish builds
its nest. It may be seen in pairs, moving in and out among
the lilies near the shore, as if jointly selecting the site for
a nursery. This is generally a gravelly spot, and once agreed
upon, the little builders vigorously commence work. The stems
or roots are torn up for several inches about, and carefully carried
to a distance away, while the smaller rootlets are swept aside by
skilful blows of their tails, both fishes often standing over the
nest creating a mimic whirlpool with their fins that effectually
carries off the objectionable particles. The stones are next taken
up, the smaller ones in their mouths, the larger being pushed out
bodily, or fanned away by the sweeping process, until finally an
oval depression appears, with a fine sandy bottom. The stems
and other aquatic verdure about the sides, that seem to have been
purposely left, now naturally fall over, so that oftentimes the nest
is a perfect bower, its walls bedecked with buds, while the roof
is a mat of white lilies floating upon the surface. Here the eggs
are deposited, the male and female alternately watching them.
We have always known the sun-fish as the most peaceful of the
finny tribe, and only in wanton playfulness chasing the golden

carp; but let a stranger, a bewhiskered cat-fish, approach the bower, and war is at once declared. The little creatures seem to snap with rage and defiance, the sharp dorsal fins stand erect, the pectorals vibrate with repressed emotion, while the convulsive movements of their powerful tails show that they are ready to stand by their homes to the last; and indeed, so vigorous is their charge, that large fishes are forced to retreat, and as the sun-fishes build in companies, the intruder is often attacked by an entire colony of them.

Nearly all the sun-fishes are nest-builders, some forming arbors, as we have seen; others, as the banded variety (*Mesogonistius chaetodon*), scoop out nests on the sandy shores, rearing their young in the late spring; while the spotted sun-fish (*Enneacanthus obesus*) is more democratic, affecting muddy streams, and, as cold weather comes on, making a nest for itself in the muddy bottom, where it lies dormant till the coming spring.

Alike as are many of the members of this family in their habits, we find that the common perch (*Perca fluviatilis*), unlike its piratical cousin, builds its nest in mid-winter, its operations having been watched by careful observers through the ice. It forms a clearing much after the manner of the sun-fish, without, however, the decoration and romantic surroundings that are the possibilities of spring.

In some quiet nook or corner we have made friends with the dace (*Rhinichthys atronasus*), another little nest-builder, and a veritable finny jester. Stretched upon the green turf that overshadows their homes, we have caught glimpses of them, and, perhaps unseen, played the spy upon their domestic doings.

Life to them is a gala time. What games and sports they

have! Now in jest they join in the chase of some intruding minnow, suddenly changing their course and rising to dash at some resplendent dragon-fly that, with staring eyes, hovers over the leafy canopy of their home; again they dart about the surface, rising at imaginary flies and bits of floating weed. One more daring than the rest fairly clears a lily-pad in its leap; another lands upon the partly submerged leaf, the momentary struggle to escape attracting the attention of the sharp-eyed king-fisher, who dashes down fiercely in fruitless chase, a dire warning to the sportive fishes. All is not play, however, even among the dace. In the warm weeks of June come the sterner duties, the nesting-time; male and female join in the preparation, and the locality is selected, perhaps in some running brook, in shallow water. Roots, snags, and leaves are carried away, both fishes sometimes tugging at a single piece, taking it downstream, and working faithfully, until we, who are watching from the bank, see a clearing over two feet in diameter. Here the first eggs are deposited, and the male who has retired soon appears from up-stream, bearing in its mouth a pebble, that is placed in the centre of the clearing. Now they both swim away, soon returning, each bearing a pebble in its mouth, that is dropped upon the eggs. Slowly the work goes on, until a layer of clean pebbles apparently covers the eggs; now the female deposits a second layer of eggs, and more pebbles are brought, the little workers scouring the neighborhood for them, seemingly piling up eggs and stones alternately until the heap attains a height of eight inches or more, formed in various shapes, sometimes pyramidal or dome-shaped—monuments of the patience of these finny housekeepers. Who would suspect their purpose? Even the gleaners of the golden fields, in whose waters our little friends are found, have not discovered their secret, and think the curious piles washings of the brook itself.

THE LAMPREY-EEL AND NEST.

Very similar to the Dace in their habits of erecting a nest are the lamprey-eels (*Petromyzon marinus*). They are common on our Eastern sea-board, living alike in salt and fresh water. In the early spring they follow, sometimes precede, the shad up the rivers, and search for safe localities in which to deposit their spawn. The same process of clearing away is seen as cited in the case of the dace. Their long bodies are bent in coils, and used in pushing aside the accumulation on the bottom, and to the uninitiated the appearance of two eels, each three feet in length, twisting and seemingly coiling about one another, would be indicative of war. The water having cleared, a smooth spot becomes visible. Upon this the lampreys proceed to place stones, (*Plate I.*), the size of some of which is almost as astonishing as the intelligence they exhibit in transporting them. Irregularly shaped stones of small size are easily and quickly brought in their mouths from the several localities in which they forage; some they are able to carry only a few feet; then, dropping them, they push them along by main force. But when stones that weigh several pounds are to be brought, they adopt tactics worthy of an engineer. As the spots in which they rear these submarine castles are generally subjected to a swift current, the largest stones that it would be thought impossible for them to move, are looked for up-stream only. A suitable one found, it is moved about until a favorable portion is presented, and to this the sucking mouth is fastened; the tail of the fish is then raised aloft, and by a convulsive effort the heavy stone is lifted from its place, the current pushing against the fish and stone, sweeping them along several feet before they sink ; another effort on the part of the fish, and the rock is again raised and carried down-stream, until finally, by repeated liftings and struggles, the ingenious nest-builder is swept down to the nest, and its load deposited. This laborious work is carried on until the pile assumes a height of two or three feet, and a diameter of four. No special shape seems to be desired, it being generally oval and compact, well devised to contain the eggs, which are deposited within, affording protection in its many interstices for the young when they hatch. Strange little fellows they are. When about six inches long, they have no teeth, are blind, and possess so many characteristics to distinguish them from the adult form, that for a long time they were considered distinct animals, and described as a different genus (*Ammocotes*).

The largest nest that has ever been observed was found by Mr. John M. Batchelder ; interesting as showing that as many as fifty eels join in building a common nest. Mr. Batchelder says : " During the month of June I had an excellent opportunity to observe the manner in which the lamprey-eel (*Petromyzon marinus*) builds a stone dam for the deposit of spawn, and for the protection of the progeny.

" The location of the structure was in the Saco River, Maine, within the ripples near the foot of the lower falls, three miles from the sea, and near the level of mean high water. It was nearly at right angles with a shore-wall of granite, and was about fifteen feet long, and from one to three feet in height. Its position and triangular shape in vertical section were well adapted for securing a change of water, and a hiding place among the stones for the young.

" When I first noticed the movements of the eels, they were

diligently at work, their system of operation being very me-
thodical; but I was not able to determine whether there was
any action by single pairs, as their movements were rapid, and
the number engaged at one time must have been fifty, while it
is probable that a hundred were at work, for they were constantly coming from various directions to take or resume their
places on the up-stream side of the dam.

"The river-bed at this point was made up of water-worn
stones, chips of granite, and fragments of bricks, over which
there was a steady flow of water, the depth being four or five
feet, but varying with the level of the tide.

"The mode of raising the material was the same in all cases:
the eel attached his mouth to a stone, and then, with many
wrigglings and contortions (the head always pointing up-stream)
lifted it from the bottom; he then backed down stream, floating
with the current, until the stone was over the centre of the heap,
when it was dropped, lodging sometimes on one side, and sometimes on the other. He then usually returned for more material
to the deep and comparatively still pool formed above the dam
by the previous excavations, but in some instances was unable
to stem the more rapid current at the top of the dam, and was
carried below it. When this happened, he swam around the
outer end of the dam, and returned to the pool to resume the
work.

"I noticed in many instances that the heavier stones were
lifted by two eels, working alongside of each other, and carried
to their proper places in the structure. Half-bricks, weighing
two pounds, were thus transported by one individual, and many
of the stones were of much greater weight. Later in the season
many of the eels were lying quietly upon the up-stream side of
the dam, and about the middle of July all had disappeared.

"The temperature of the water, when the river-current was not met by the tide, was in June about 64° F., and in July, 71°.

"Stones of various sizes, lying at the base of the shore-wall, were removed; and it was evident that the stability of this wall would have been impaired if it had been built upon a pebble or gravel foundation, instead of a solid ledge."

The most remarkable work performed by fishes, as far as my observations go, are the nests of the fresh water chub, *Semotilus bullaris* known in some localities as the stone toter, a fish attaining a length of fifteen inches or so. The finest nests that I have found are on the shores of Westminster Island, and they are common on nearly every island that has a sandy, gravelly shore among the many that make up the assemblage known as the Thousand Islands. The first one I saw was on the edge of the little channel known as the Rift, and it was pointed out as an ash-dump from a steamer, standing in four feet of water, rising sufficiently near the surface to obstruct a sail boat. The largest heap that I have examined was about ten feet across at the base and four feet high, approaching within a foot of the surface, and contained a large cart-load of stones weighing in all perhaps a ton. The stones ranged from small pebbles to some four inches long, and as some of the nests were quite a distance from gravel beds each stone represented a journey, and as there are tens of thousands of them, the amount of labor performed by these finny workers can be imagined. Each stone is brought in the mouth of the chub and dropped over the pile, one or more fishes working at the same heap. It is possible that there is a rude plan followed in the work, as in a number of nests evidently recently commenced the first deposit of stones was small, and dropped to form a circle or semicircle.

PLATE II.

GOURAMI AND NEST.

The largest heaps are undoubtedly the work of successive years, the fishes piling up the stones year after year just as the bird Megapodius heaps up leaves and other material in the same place on successive seasons. Again, the stones on top were always fresh, as if in endeavoring to carry them the alga or moss had been worn off, while those forming the base were moss-covered. On many of the nests I readily reached the top stones from the boat, and one of the largest weighed nearly four ounces; undoubtedly the fish carry stones much heavier. The nests are built or added to during the last of May or June, and at this time the chubs are seen lying on the heaps, when the eggs are probably deposited. All the labor of piling up is to protect them from predatory fishes, a necessary provision, as cat-fish, rock-bass, perch, and others prey upon the eggs. Whether the latter are special delicacies or not they are well protected, washing into the crevices and interstices probably as soon as deposited, remaining until hatched.

The trout excavates a simple nest in gravelly beds not incomparable to the nest of some gulls, that is a mere depression in the sand. The nest of the salmon (*Salmo salar*) is a furrow in the gravelly bottom, often ten feet in length, the depression being made as fast as it is required, the fish forming it with its tail, and covering up the eggs in the same way. In the Canadian rivers these nests can be distinguished by the lighter marking on the bottom.

Who of those fond of idly drifting along our sea-shores in admiration of the panorama below, are not familiar with the quaint toad-fish (*Batrachus*), that in its shape and color so closely mimics a moss-covered stone, finding in this resemblance an effective protection against its enemies. The parent fish intrenches itself among the weed and gravel carelessly

thrown aside, after the fashion of some of the gulls, and here
the young are reared, their yolk sacs enabling them to cling to
the rocks of the nest soon after birth, where, under the watchful eye of the parent, they remain until old enough to swim
away.

In some neighboring stream that sooner or later finds its way
to the sea, we shall find the most vigilant of all nest-builders,
the four-spined stickleback (*Apeltes quadracus*). The different
species, though very similar in their general architectural ideas,
vary mainly as to location. Some place the nests upon the
bottom, concealed among the wrack that abounds there; others
are hung pendent from some projecting ledge, or swing in the
tide from the sunken bough of some overhanging tree, there
undergoing a motion akin to rocking.

The work of nidification is performed solely by the male
stickleback, the female taking no part in the labor, and when
the spawning season arrives, he, having assumed a bright nuptial
lustre, shows extraordinary activity in securing a site for his
edifice, and transporting the building materials thither. These
are fragments of plants of all kinds, which he often seeks at a
distance and brings home in his mouth. He arranges them so
as to form a kind of carpet work, but as there is some danger
of the current carrying away the light materials, he brings sand
to weigh them down and keep them in their places. Then,
having entwined them with his mouth to his satisfaction, he
slides gently over them on his belly with a vibratory motion of
the body, and glues them together with the mucus that exudes
through his pores. Having in this manner firmly established
the floor of his edifice, he seeks somewhat more solid materials
for the walls—sometimes bits of wood, sometimes pieces of
straw—which he always seizes with his mouth, and lays either

on the surface of the floor or sticks into its sides, withdrawing them and thrusting them in anew until he is satisfied ; or, if he cannot adapt a piece properly to his building, he carries it to some distance from the nest and rejects it. After the side walls are erected the little architect proceeds to throw over the chamber a roof of the same materials with the floor, and to give firmness to the whole structure he again and again creeps over it, and by the rapid action of his fins and the vibratory movements of his tail fans out the light and useless particles. In carrying on his building operations he takes care to preserve a circular opening into the chamber, often thrusting in his head and a great part of his body, widening and consolidating it so as to render it a fit receptacle for the female. When choosing material the fish has been seen trying its specific gravity by letting it sink once or twice in the water, and if its descent was not rapid enough finally abandoning it.

The most remarkable part of this building, that has been discovered since Costa made the above observations, is the exact method by which the fish binds the nest with cords, keeping it in shape. The investigations that led to the discovery of this were made by Prof. Ryder, of Johns Hopkins University. He found that in the four-spined stickleback the male-fish binds them together by means of a compound thread, which he spins from a pore or pores, while he uses his bobbin-shaped body to insinuate himself through the interstices through which he carries the thread with which he binds a few stalks of *Anacharis,* or other water weeds, together, bringing in his mouth every now and then a contribution of some sort in the shape of a bit of a dead plant or other object, which he binds into the little cradle in which the young are to be hatched. The thread is spun fitfully, not continuously. He will go round and round the nest for perhaps a dozen times, when he will rest awhile and begin again, or turn suddenly and force his snout into the top with a vigorous, plunging motion, as if to get it into the proper form. Its shape is somewhat conical before completion, an opening remaining at the top through which it is supposed he introduces the eggs. The thread is wound round and round the nest in a horizontal direction, and if this is placed under a microscope when freshly spun, it is found to be composed of very thin, transparent fibres to the number of six or eight, where they are broken off they have alternated tapering ends, as though the material of which they were made had been exhausted when the spinning ceased. Very soon after the thread is spun particles of dirt adhere to it and render it difficult to interpret its character. Prof. Ryder saw the thread drawn from the abdomen repeatedly, and it appeared to him probable that it came from the openings of a special spinning gland. The nest measures half an inch in height and three-eighths in diameter.

The time occupied in collecting building materials and constructing the nest is about four hours, and when all is ready the male proceeds to seek a female, and, having found her, conducts her with many polite attentions to the prepared home. The female of another species, according to Sir John Richardson, enters the nest by one door, and, having laid several eggs, escapes by the opposite outlet, leaving the eggs exposed to the current of cool water which flows through the nest. Then the male establishes himself as guardian of the precious deposit, not suffering even the female to approach it again. Every fish that comes near, even though much larger than himself, is furiously attacked; and he gives battle valiantly, striking at their eyes and seizing their fins with his mouth. His acute dorsal and ventral spines are effective weapons in these combats. The constant watchfulness of the male is needed; for, if he is removed by way of experiment, the sticklebacks and other fish lurking in the vicinity rush with one accord upon the nest, and devour the eggs in an instant. For a whole month does the male provide for the safety of his offspring. In the first few days the openings are enlarged so as to admit a larger current of water to the eggs; and about the tenth day the male employs himself in tearing down the nest and transporting the material to some little distance. With a lens the fry at this time may be observed in motion. Around these the male guardian continually moves, suffering no encroachment; and as the young brood gain strength and show an inclination to stray beyond bounds, he drives them back within their precincts, until they are advanced enough to provide for themselves, when both old and young disappear from the place of observation.

In specimens kept and observed by me the male assumed a fiery-blood-red hue at this time, and was constantly on guard, dashing at my hand if presented at the glass, even injuring himself by the violence of his blows. This vigilance was kept up until the eggs hatched, and then when the young offered to stray, the father-fish would draw them into his mouth and shoot them out violently in the direction of the nest. As they grew larger and stronger, and strayed away to a greater distance, his patience became exhausted, and one day he suddenly deserted them.

THE GREAT SARGASSO SEA.

In the vast area of floating sea-weeds, occupying 260,000 square miles of surface in the Atlantic and known as the Sargasso Sea, are found numbers of animals that seem peculiarly adapted by various modifications to the pelagic life they lead.

On the outskirts of this sea great detached patches of *Fucus* and *Sargassum* are often found available to the voyager, richly repaying a passing examination of the nomadic inhabitants.

During many hours spent in drifting in the Florida Straits, surrounded by similar thickly growing and matted weed of the Gulf Stream, I renewed acquaintance with one of the quaintest and most skilful of the marine nest-builders.

Collecting in the gulf weed requires no little prescience, as the inhabitants, one and all, from the soft shell-less mollusk *Scyllœa*,* to the short-tailed crab *Nautilograptus*, have assumed the exact tint of the surrounding weed—a protective resemblance that serves them well—but a close examination soon reveals myriads of strange creatures. Though the strong trade-wind is blowing, the great patches of weed are so profusely distributed, that the intervening stretches of clear blue water are smooth as glass. Now the claws of some quaint crab wave a moment in the air as it essays a submerged bunch of weed; dazzling forms of gurnards, with their lace-like wings and burnished helmets, rise and soar away over the grassy sea. The warm wind is burdened with saline odors; the blue channels among the weeds scintillate with golden reflections; and from far away comes the weird "Ha! ha!" of the laughing gull, that, with the occasional splash of the pelican, is the only sound to be heard in this ocean world.

In the full enjoyment of our novel surroundings, I was attracted by a singular object peering out of the water; the boat being pushed nearer, the curious creature proved to be the pelagic fish, *Antennarius marmoratus,* so exact in its imitation to the

---

* In many instances there is no common name for the object mentioned. In such cases the technical terms can only be used. Those who desire to have more definite information can find it under the respective heads in text-books.

gulf-weed (*Sargassum*), that, had I not been familiar with it, it would have been passed by. The tall and barbeled dorsal fins were out of water, as well as the curious seeming horns that decorate the head, and thus, half submerged, the little fellow appeared about to take leave of its native element, and walk away over the weed. It was resting upon its nest, an oval ball of *Sargassum*, a little larger than an Edam cheese. This curious creature, whose pectoral fins resemble limbs, selects from the floating algæ bits of *Sargassum bacceferum*, which consists of feathery bunches, each tuft having a thread-like branching stem studded with round air-vesicles that form perfect floats or buoys. These are collected into a single mass by the fish, and woven in and out in a seemingly incomprehensible manner. A bit is taken in its mouth, with which the fish dives into the mass, coming out at the opposite side. As the nest assumes a more compact shape, a gelatinous substance is attached to the various parts that serves to connect them. It is now an irregular oval, floated by the natural buoys. Now the eggs are deposited, and attached to the weed by some secretion. This done, other pieces are added to partly conceal them, and the fish passes repeatedly around the nest, rubbing its abdomen against it, and binding it together by silken bands of a visceral secretion that it takes, perhaps, from certain glands, as in the case of the stickleback. This completed, the strange inhabitants of this pelagic world lend their aid in its adornment. The rich bryozoon (*Membranipora*) incrusts the various parts with its silvery growth, the nest itself throwing out new shoots, their tips assuming rare tints of yellow and green, in strong contrast to the dark shades of the older forms. Graceful stalked vases of the *Campanularia* appear as if by magic; small barnacles hang pendent upon the leaves, while delicate shapes of *Ianthina*, *Vellela*, and *Porpita*, glistening in blue and silver, with the fantastic *Glaucus* and luminous *Salpa*, hover about in close attendance. Around the nest the quaint parents move, or recline upon it, as we have seen. When the small eggs hatch, the bands become loosened, and in the nest, that is a veritable living arbor, the young find abundant protection, and closely resemble the bits of weed among which they lie concealed.

Among the fantastic gobies are several that vie with the birds in nest-building. In the great submarine tangles of the Mediterranean Sea, where grim kelps wave their long-leaved stalks, the black goby (*Gobius niger*), according to Olivi, builds its home. The finer bits of weed, *Zostera* and others, are bound and interwoven in irregular form, and in the nest are placed the eggs. As with the stickleback, it is the male that erects the nest, and, after the eggs have been placed within, mounts guard, remaining in watchful surveillance long after the young are apparently large enough to take care of themselves.

THE GOURAMI AND NEST.

One of the most interesting nest-builders is the famous gourami, (*Osphromenus goramy*), originally a native of China, but introduced into many countries. According to Baron de Roujoux, the gouramis attain in their native country a length of six feet, and weigh over one hundred pounds. But their nest-building is the most remarkable, showing that they rank with the land animals in intelligence. In their family relations, according to Gill, they resemble the sun-fishes of temperate North America, and the cichlids of tropical America and Africa. The spawning season, according to this author, falls in the autumn (March and April) and spring (September and October) of the transequatorial

PLATE III.

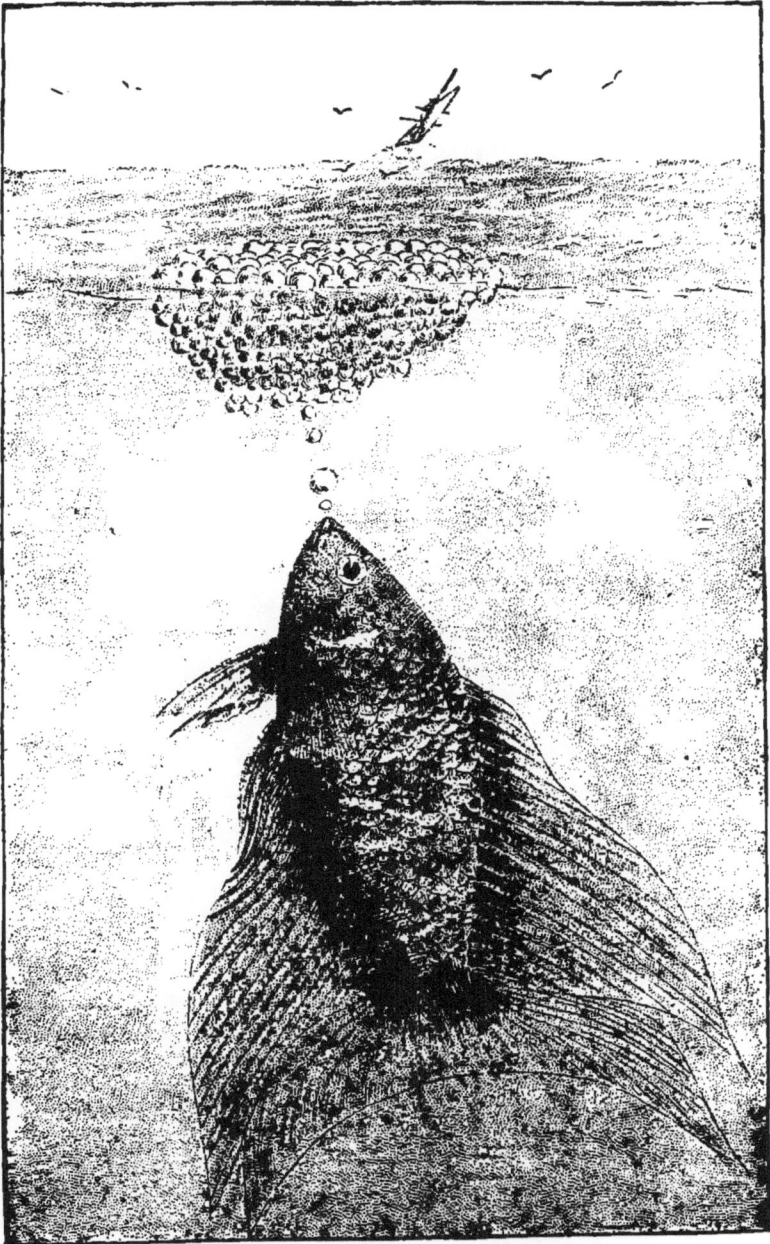

PARADISE FISH BUILDING ITS BUBBLE NEST.

islands of Mauritius and Bourbon. When that time has come the males and females pair off, and each pair select a suitable place wherein they construct a rude nest. Like all intelligent animals, they will only propagate when insured a suitable temperature for the eggs and young, a fit retreat for the building of the nest, with plants and mud for its construction, and aquatic plants for the food of the young. The bottom selected is muddy, the depth variable within a narrow area; that is, in one place about a yard, and near by several yards deep. They prefer to use for the nest tufts of a peculiar grass, (*Panicum jumentorum*), which grows on the surface of the water, and whose floating roots, which rise and fall with the movements of the waters, form natural galleries, under which the fish conceal themselves. In one of the corners of the pond, among the plants which grow there, the gouramis attach their nest, which is of a nearly spherical form, and composed of plants and mud, and considerably resembles in form those of some birds. (*Plate II.*)

The nests, of course, vary in size in proportion to the fishes, but the usual size is somewhat less than a man's hand in length. The fishes are employed some five or six days or a week in building, and their task is rendered easier, when the pairing season has arrived, by placing in the water, almost at the surface, branches of bamboo, to which are attached bundles of fine dogs'-tooth grass. The gouramis take this grass, and with it form their nests in the branches of the submerged bamboo, in a manner analogous to that with which the common silk-worm avails itself of the branch which is presented to it to make its nest on. When the nest is wholly completed the female deposits her eggs, which in a moderate-sized individual amount to about eight hundred to one thousand. After the eggs have been deposited and while they are hatching, the parents remain

2

near, jealously guarding them and rushing with vehement fury
at any ordinary intruder near their domains, and thus they
continue to guard the young for a time after they are hatched.
The eggs are soon hatched, and in the nest the young find a
refuge where they are free from a thousand dangers to which
they would be otherwise exposed during the first days of their
life; and in the macerated vegetable matter of which their home
is partly composed they obtain their earliest food, and that
which is most suitable to them in their most delicate condition.
Soon, however, they make short excursions from the nest though
under the guidance of the parent-fish, who is prepared to give
them aid in case of need. They do not soon disperse, but keep
together in shoals.

THE PARADISE FISH AND ITS NEST.

In Siam there is found a fish called the plakat, known to
science as the *Macropodus* or paradise fish, on account of its
curiously shaped fins. These fishes are kept by the Siamese as
pets, being carried about in jars of water and fed upon the larvæ
of mosquitoes, and trained to fight for the amusement of their
owners. When the paradise fish builds its nest, it uses no hard
material, the male merely rising to the surface and sucking in
air, carrying it down, issuing it as a mucus-covered bubble.
This is done an innumerable number of times, until repeated
layers of them are deposited, forming a quite large mass of the
filmlike bubbles of spittle. This floating mass constitutes the
nest, and in it the eggs of the female are deposited; and, wonderful to relate, the material of which the nest is made forms the
first food of the young fry, and by the time they have devoured
the nest so that they will drop out they are large enough to take

care of themselves. Does this recall the bees in the method of providing for their young?

Those observed by myself were in confinement in the museum over Fulton Market, New York, having been brought from India by the captain of a sailing vessel. I was particularly struck with their seeming intelligence, if this term can be applied, or perhaps vivaciousness would better describe it. Never at rest, they were continually on the alert, darting about with graceful movement, trailing their beautiful plume-like fins behind. At any movement of my hands over the water, they would rise without fear and follow the finger in play; darting around as if ready for any game or sport that might be suggested. The affection of the pair was particularly evident. In approaching each other they would stand in mid-water, face to face, and move round and round, retaining the same position, as if in the maze of some mystic dance. At this time the broad fins of the upper and caudal portions would be fully extended, and fairly vibrating as if with emotion, presenting a striking appearance. As soon as the fish separated the fin rays dropped, seemingly reducing the size of the animal one third. I also noticed that whenever the two met, they expanded the opercula, or gill-covers, so that they stood out very distinctly, showing the gills within, an operation that reminded me of the pouting of monkeys when engaged in an interchange of abuse, though the action on the part of the paradise fishes was evidently a friendly one, as they made no attempt to attack each other.

The attempts at nest-building were very much the same as previously described. The male would rise to the surface, as he often did in play, readily taking objects from its owner's hand, and inhale pure air, then sinking beneath the surface would somewhat forcibly emit the supply of oxygen in the form of bubbles, the covering of which was undoubtedly a film of mucus. These rapidly rose to the surface and clung closely together. (*Plate III.*) By repeating this an indefinite number of times an area of bubbles was formed three or four inches in diameter, the continual supply having the effect of lifting the upper layer out of water, so that the small portion of the nest exposed to the air assumed a slightly convex shape, like the crystal of a watch. When complete, or nearly so, the nest was several inches deep, very fragile, and easily blown apart, but probably age and fungous growth would soon render it more tenacious.

His work ended, the male began to chase his mate around the aquarium, as if to drive her toward the nest.

Professor Rice, to whom these fishes belong, is of the opinion that he could distinguish a space, or air-chamber, between the upper layer of cells and those lying directly upon the water, and suggested that the eggs might be deposited there or find their way to it. These fish are of a rich olive hue, marked with oval spots, lacking, however, the coloring that distinguishes their allies of the Eastern and tropical seas.

The famous *Serrasalmo* or perai, of South American waters, often selects for its nest a structure that reminds us of the home of the oriole. On the borders of the rivers great vines or *lianes* climb the palms, and creeping out upon the leaves finally fall down until they reach the water. There rootlets grow, foreign matter is collected, and finally a veritable miniature floating garden swings in the current. Beneath this arbor, amid the clustering roots, the perai is wont to place its eggs, and there they are watched and guarded until the fry can safely leave their swinging cradle.

The cat-fishes of this and neighboring regions are noted for their domestic habits. The armored *Callichthys* (*Plate VIII.*), according to Schomburgk, lays its eggs on straws "which they gather together and cover over. They are watched by the parent until they arrive at maturity. The adults are taken in trenches at this time by putting a basket near the nest, with which it is lifted out of the water."

The most remarkable statement concerning these fishes made by this author is that the Hassar, which is the native name for *Callichthys subulatus*, constructs a regular nest of blades of grass and leaves in holes *just above* the water, where it deposits its eggs, and watches them until the young appear.

The great Ceratodus of Australian rivers forms a rude nest in the weeds at the bottom of the stream. The lump-fish is also accredited with being a nest-builder, and, unlike many others, the young follow the parent fish. Fishermen call them hen and chickens at that time. The young have the curious ventral suckers that characterize the parent, and according to some naturalists, cling to the male. In fact, in nearly all these instances of affection among our finny friends, the mother runs away, leaving the father to perform all the domestic duties. The Solenostoma and Aspredo are the only cases among fishes in which the mother does not desert her young.

CHAPTER II.

LIVING HOMES.

IN glancing at the nest-builders of the sea, we have observed only one side of the home-life of our humble friends. We may term the former the educated classes, the architects; but just as among the human inhabitants of the world, there is a class whose members are dependent upon others. They do not seem to have the faculty of caring for themselves, and so find their protection in the companionship of other forms. In fact, they have living homes, taking up their residence with or in the person of some larger animal. They must not be mistaken for parasites, that prey upon animals they live with. They may at times overstep the bounds of hospitality, but in the main they merely use their living refuge as a dwelling-place, and in many cases become so dependent upon it that they never stray into the outer world.

During several years' residence on the growing atoll of Tortugas, on the Florida Reef, many opportunities were afforded for observing instances of this peculiar phase of existence. The group, comprising seven or eight keys,* made up of coral, is surrounded by a long reef that almost completely skirts the latter, composed of dead coral heads, against which the sea beats powerfully. A few feet within this range of breakers the reef stretches away; a field, as it were, of pure

*The term "Key" which will be repeated often, is a corruption of the Spanish Cayo, an islet.

22

white sand, covered here and there with patches of corals of many kinds.

Here was the collector's paradise, as strange creatures, crabs, fishes, and shells, appeared at every move, and dropped in showers from every piece of coral lifted from the water.

One of the commonest animals in this submarine garden was a long, worm-like creature, called the trepang; or, scientifically, *Holothuria*. It looked like a great black cucumber lying on the bottom, and was so tough that with difficulty could a spear be sent into its leathery hide. I was wading along in about three feet of water one day, towing the boat, when wishing to secure one, I picked up a specimen, ten inches perhaps, in length. The curious creature slowly doubled up when taken from its native element, and lazily ejected a stream of water from its mouth as it was placed in a glass jar, carried for aquarium purposes. In a short time the holothurian exhausted the air in the water of the jar, and began to move about restlessly, and soon something else was affected by the lack of aeration, as a beautiful, silvery transparent fish was seen thrusting its head from the body of the holothurian, and in a moment fairly made its way into the outer world. It moved slowly about for a short time, making ineffectual efforts to swim, and though the water was immediately renewed, it soon died.

The fish was about six inches in length, and evidently a creature unused to the outside world; a veritable phantom fish, so ghostly white and transparent that, if laid over a newspaper, print could be read through it. In general appearance it resembled an eel; its upper and lower fins were all joined, and extended the entire length of the body. Its scientific name is *Fierasfer dubius*, and nearly all the family take up their abode in other animals. Many of these curious boarders were found afterward; in fact, very few of the holothurians were without them, but, in every case, though the greatest care was taken to have the water fresh, the fish died when exposed to light and open water; and the supposition was that here, at least, certain individuals never left their host.

Since then another holothurian has been watched, in the Naples aquarium, and found to afford shelter to a fish that comes out and returns at pleasure, the latter operation being performed in a most remarkable manner. The curious sea-cucumber that is thus converted into a boarding-house, forces water in and out of its body at short intervals, so that there may be said to be an intermittent current running in and out. When the fish wishes to return it takes advantage of this, and inserts its tail in the orifice that forms the door of its house, and as the animal takes in water the fierasfer is drawn along with it, and by degrees gradually reaches the interior. That it goes in tail first would seem a singular performance, but being a long fish it could not easily turn if once in head first, so by entering in the opposite direction it is always ready to come out again.

These cases are not confined to certain localities, but all the fierasfers of the world seem inclined to adopt the sea-cucumber as a home. At Madeira, Dr. Greef found one occupied in the same way, and Quoy and Gaimard state that a fierasfer is a dweller in *Stichopus tuberculosus.* As the holothurians are also tenanted by various shells and crabs, they not inaptly suggest the boarding-house.*

Though many of the fishes have been examined, there is yet

---

* In the Agassiz Museum, Cambridge, Mass., is a valve of a pearl oyster, in which a *Fierasfer* is beautifully enclosed by a pearly deposit. The fish being a permanent boarder in the living oyster, the latter impatient at the intrusion covered it as by the same process that forms valuable pearls.

PLATE IV.

HERMIT CRAB AND ITS BOARDER AS IT APPEARS UNDER WATER.

much discussion as to their method of obtaining a living. Some naturalists contend that the fierasfer goes without to obtain its dinner; others, like Semper, have it, that when hard pressed the fish will prey upon the body of its protector; indeed, the latter naturalist, in the Philippine Islands, opened a number of fierasfers that he took from sea-cucumbers, and found there decided evidence that they were living upon the respiratory processes of their friend, that fortunately was able to reproduce them, and was not injured in the slightest.

The fierasfers are not the only fishes that take possession of these much-abused animals. At Zamboanga there has been found a species (*Scabra*) in whose stomach there lives a little fish, of a totally different genus (*Enchelyophis*), but whether it comes out after the fashion of the Naples fierasfer or not is not known.

The great mouthed angler (*Lophius*), found on our coast, offers a retreat in its mouth for a number of fishes that use the cavity for a house. At Nice, an allied form, that is there called the *Beaudroie*, gives lodging in its bronchial sac and gills to a small eel-like fish of the family *Murenidæ*, a number of little crabs also sharing its home.

The beautiful sea-anemone is often made the victim of some playful fish, remarkable from the fact that the former has a terrible armament, being covered in many parts by lasso-cells that hurl out sharp, poisonous darts at the slightest warning. In Chinese waters an anemone about two feet in diameter is nearly always thus tenanted; the little fish, when alarmed, darting away in the direction of its protector, that whether by a mutual understanding or from fright, immediately closes its mouth and perfectly conceals the tenant, perhaps to the wonder and astonishment of the follower.

Lieutenant de Crispigny, an officer in the French service,

kept an *Actinia crassicornis* for a long time in an aquarium. One day, soon after securing the creature, he was surprised to find a fish in the globe. When he attempted to observe it, it darted around as if greatly alarmed, and finally made a leap, like the harlequin in the play, and landed fairly in the centre of the anemone and disappeared; the latter closing its tentacles over it. These pets were kept by the officer for over a year, and he became thoroughly convinced that here was a positive case of friendship between the totally different creatures.

Even more wonderful yet is the fact that star-fishes offer homes for fishes. One, in the Indian Ocean, known as the *Asterias discoida*, is tenanted by a little fish, but whether it often leaves its host, presumably a very difficult operation, has never been determined.

Fishes of the great mackerel family are remarkable for their associations; thus many, when young, live up among the tentacles of the great jelly-fishes; and among numerous jellies that I have examined in tropical waters, rarely was one found that did not afford protection to one or more finny dependents.

One of the most interesting cases is observed in the physalia. This animal is a colony, the individuals of which are called zooids. Some are for locomotion, others feeders, while others again are for various purposes; but to the casual observer, they all look alike—a blue, entangled, jelly-like mass of threads, coiling, drawing up and lowering again, like so many snakes, floated and suspended from a rich balloon-like bubble. On closer examination the secret of their power is seen. The tentacles are covered with delicate cells, each of which contains a minute coiled dart, called a lasso, and the moment a foreign substance comes in contact with the tentacles, millions of these weapons are hurled into it, like shots from a gun, producing a peculiar paralyzing effect. This is particularly noticeable in fishes, and I have repeatedly observed the tragedy. The long, tempting tentacles, streaming invitingly down, would be seized by a small sardine; a wild leap of fear, perhaps, and the victim would roll over dead, to become entangled and absorbed by the living man of war.

The effect of the darts upon human beings is almost equally fatal, if contact is made at the right spot. I have been stung a number of times by them upon the hand, when the pain resembled that occasioned by scalding water; but upon one occasion, in swimming upon my side, I passed completely over the tentacles of a large physalia. If I had been in deep water, nothing could have saved me, as my chest and abdomen were covered with the blue mass that seemed to penetrate into the flesh like molten lead. I regained my feet, but could move only with difficulty. Shortness of breath was, perhaps, the most serious symptom, which with the terrible burning rendered me almost powerless. Some workmen near by plunged into the water, and carried me to the shore, where the blue mass was scraped off with knives. Applications of oil and generous use of stimulants acted favorably, but the virulent nature of the poison can be judged from the fact that for over six months I could have passed as a very respectable tattooed man, the flesh being covered with fanciful blue tracings resembling designs in India ink.

On one occasion, while rowing along the surf on the reef, I found a hawk-bill turtle lying on the surface, completely paralyzed by a physalia, or Portuguese man-of-war, as it is called, scarcely larger than a hen's egg. The blue tentacles had entirely encompassed the animal's head, and the darts or lassos had penetrated the unprotected skin around the eyes.

I have thus described the power of the physalia to more clearly show the wondrous nature of the friendship (?) that exists between it and several small fishes, of the genus *Nomeus*. Up among the deadly lobes of this fairy ship, they are found swimming about perfectly fearless, dodging in and out among the tentacles that are instant death to others, and evidently aware that their protector is little likely to be attacked by other animals. Perhaps the most remarkable part of it is that the fishes are a vivid blue color ; in fact, the exact tint of the tentacular part of their host ; so that they are not only protected by it, but are likely to be taken for the tentacles themselves.

In Brazil is found a curious fish, allied to our cat-fishes, that in its mouth gives shelter to a number of small fishes that lodge in the various cavities, and seem to find perfect protection. For a long time these were supposed to be the rightful progeny of the *Platystoma*, but examination showed them to be fishes of an entirely different kind, and full-grown, though small.

The pilot-fish that accompanies the shark, in one sense makes its huge companion a home ; at least, finding protection in the companionship.

Another companion of the shark, however, frequently shares its fate. This is the remora, that bears upon its head an oblong plate or disk, arranged with transverse rows, like the slats on a blind. When the remora presses this upon the shark, the air is exhausted, and thus it clings, so that it may be said to live upon its huge companion. I have frequently hauled sharks high and dry, and had to pull the remoras off by main force to secure them.

Among the crabs there are many that constitute homes for other animals, while many others have living homes themselves. Of the latter the *Pinnotheres* is the best known. One species is the little crab we find in oysters, and from their habit of forcing themselves upon the oyster, making a home among its soft folds, they share its fate. The association, however, redounds to the benefit of both, as the crab may possibly drag in bits of food for the oyster to eat, while the latter affords it protection from the predaceous fishes. They inhabit a great variety of shells, and, as we have seen, the air-chambers of the sea-cucumber. In Northern waters a shell known as the *Modiola papuana* affords a home to two of these crabs that attain the size of a chestnut, and · it has been impossible, so far, though hundreds of the shells have been opened, to find one without its quota of crabs.

The large shell Avicula, the one that affords the best pearls, is the home of a *Pinnotheres,* also of a crustacean allied to the lobster, and perhaps the material dragged in by these boarders has formed the base for pearls that have sold for thousands of dollars.

In Eastern waters is found the giant clam *Tridacna*, that often weighs five hundred pounds, the animal alone thirty. The largest shells are nearly five feet in length, and are often used for ornamental purposes, or as a play-house for native children, as shown in the accompanying picture or frontispiece. When alive the great clam gives shelter in its folds to a number of crabs, one especially, known to naturalists as the *Ostracotheres tridacnæ,* nearly always being found in the huge mollusk. In the pearl mussels, common on the coast of Mozambique, three crabs are found. In the Atlantic waters a large acephalous mollusk affords protection to a number of crabs, and in the *Pinna marina,* a beautiful fan shell, a crustacean, of a pale rose color, lives.

Even Pliny was aware of these living homes, and ascribes to the dwellers a mistaken office. He says the Chama is a clumsy

animal without eyes, which opens its valves and attracts other
fishes, which enter without mistrust, and begin to take their
pastime in their new abode. The pinnothere, seeing his dwell-
ing invaded by strangers, pinches his host, who immediately
closes his valves, and kills one after another of these pre-
sumptuous visitors, that he may eat them at leisure.

Rumphius, the Dutch naturalist, had similar ideas, and said
that these crustaceans inhabit always two kinds of shell-fish,
the *Pinna,* and the *Chama squamata.* According to him, when
these mollusks have attained their growth, one pinnothere (one
only, at least, in the Chama), lives in their interior, and does
not abandon its lodging till the death of its host. He regards
this crustacean as a faithful guardian, fulfilling the duties of a
doorkeeper.

On the Peruvian coast we find a curious little crab, that instead of taking up its abode in a shell, chooses an *Echinus* as its
home ; and how it manages to make its way through the grind-
ing teeth without getting squeezed to death, is somewhat of a
mystery.

A CRAB AND ITS BOARDER.

Around Marseilles and in our own waters a crab is often
taken by the fishermen, that carries about an anemone on its
back. The fortunate anemone, moreover, is so placed that its
mouth is always opposite that of the crab, thus receiving all the
morsels that fall from its host's table.

Colonel Stuart Wortley has paid great attention to the alli-
ance that exists between the soldier crab and the sea-anemone
Adamsea, and the movements of the crab to protect its tenant
are certainly governed by intelligence. The best morsels are
offered it, and when the crab finds that it must leave its shell, it

strives with the greatest care to also remove its friend, and by
delicately prying off its disk, ultimately succeeds in doing so.
If the *Adamsea* is not suited, and will not retain its position,
the obliging crab tries other shells until one is found satisfactory
in every respect to this curious creature. (*Plate IV.*)

One of the most astonishing cases of living homes was observed by Dr. Richter in two crabs of the family *Polydectinæ*,
the members of which have their front claws armed with large
teeth. Latreille remarked that a gummy substance was generally to be found at the ends of the claws, and Professor Dana
described the animal as having always something spongy in its
hands. Dr. Möbius discovered the remarkable fact that these
things, held in the two claws of the crab, are in reality living
sea-anemones. These anemones are attached to the immovable
joint of each claw, whilst the teeth of the movable joint of the
claw are kept buried deep into the flesh of the sea-anemones, and
thus hold them fast; although each anemone can easily be pulled
away from its position with the forceps, in specimens preserved
in spirits. The mouth of the anemone is always turned away
from the crab. The same curious association exists in the case
of another species of the same family, but of a different genus,
which also inhabits Mauritius. Dr. Möbius gives the following account of the matter: "I collected about fifty male and
female specimens of *Melia tesselata*. All of these held in each
claw an *Actinia prehensa*. The recurved hooks of the inner
margins of the claw-joints of the crab are particularly well
adapted to hold the actinias fast. I never succeeded in driving
the living actinias out without injuring them. If I left the
fragments of them when pulled out lying in the vessel in which
the *Melia* was, the crab collected them again in its clutch in a
short time. If I cut the actinias in pieces with the scissors, I

found them all again in the claws of the crab after a few hours. It is very probable that the actinias aid the crab in catching its prey by means of their thread-cells, and that the actinias, on the other hand, gain by being carried from place to place by the crab, and thus brought into contact with more animals, which can serve as food to them, than they would if stationary. This is a very interesting case of commensalism."

From all these instances there is one inference which we draw, namely, that lowly as are the animals in the scale of intelligence, we certainly must accord them the faculty of distinguishing between friend and foe, and admit that they have feelings more or less comparable to our own.

PLATE V.

PERIOPHTHALMI SEARCHING AFTER FOOD ON DRY LAND.

/

CHAPTER III.

DRY LAND FISHES.

THAT some fishes should leave the water and travel overland is, perhaps, not more remarkable than the fact that certain birds without special modification, as the ouzel, should leave their natural element and fly into and under the water. Who knows the secret paths of the great marsh but has watched the brown-hued eels wriggling their way from one pool to another through the grass, especially at night, leaving their homes and wandering about.

I have seen the great armored gar rise again and again for the air that would seem necessary to its existence, and many other fishes are equally dependent upon it. In a great number, however, there appears to be a special modification of structure, enabling them to remain for a greater or less time entirely out of water. It has long been known that the blennies are frequently left by the tide in the rocky pools of the coast, often being found under the damp weed, entirely out of water, the occurrence supposed to be involuntary; but late information tends to show that these fishes intentionally place themselves so that they will be left by the tide—a most unfish-like operation, certainly.

The most remarkable instance of this has been observed in the fish known as *Blennius pholis*. On placing a specimen in a glass vessel of sea water it appeared perfectly quiet for some hours, but at length became restless, and made frequent attempts to throw itself out of the water. It then occurred to the observer,

3   33

a Mr. Ross, of England, that, on a former occasion, when occupied at the sea-side, he had a blenny in a vessel with some *Actiniæ* and *Serpulæ*, which regularly passed a portion of its time on a stone; he therefore placed a pebble in the glass, the fish immediately leaping on it completely out of water. It thus appeared that these changes of element were necessary to its existence. On going to the front of the house the naturalist perceived that it was near low water, and knowing that it would flow until ten o'clock that night, he watched the movements of his little captive, and as the clock struck, had the gratification of seeing it plunge again into its natural element, and for over five months this remarkable fish was an accurate tide-indicator. It was noticed that the fish had the power of altering its position on the stone with great facility by means of its pectoral and ventral fins. At times it reclined on its side, at others was perfectly erect, resting on its broad pectorals, and turning itself from side to side. It took crumbs of bread and small earth-worms, two or three a day being sufficient, and became so familiar as to take its food from its owner's hand, and if not attended to, dashed the water about to let him know that he was on the look-out for his bit of meat or rice.

While in the water the colors of this blenny are less strongly marked; but after being a short time exposed to and inhaling atmospheric air, the color changes to a deeper brown, and the markings become nearly black, with a regular series of white spots above and following the course of the lateral line.

The New Zealand gobies run along the sands at low tide, much after the fashion of water-birds, jumping at the half-buried crustaceans, and moving so rapidly that they are called by the natives "running fish." The sunghong is often seen running in groups along the paddy grounds of Whampoa; while the pakkop, or white frog of China, that is carried about the streets of Canton alive, often during the barter that precedes the sale, attempts to hop away.

The flower-fish, hawaya, so commonly portrayed on Chinese ware, is an amphibious goby, that spends certainly half its time above water on land, where they are often caught. The Periopthalmus is equally at home on land, and is chased along the shore by natives, like frogs, the fish jumping from rock to rock, and not taking to the water until closely pressed. (*Plate V.*)

The habits of these remarkable fishes that form a genus of the Goby family are most peculiar, and are an enigma that scientific men find difficult to solve. A gentleman connected with the *Challenger* expedition was fortunate in observing them on shore, and expressed his opinion that they are more at home on land than in the water. He saw hundreds of them high and dry darting around as nimbly as frogs, raising themselves on the two pectoral fins, and looking around with their prominent eyes in a most comical manner; but it was found extremely difficult to catch them.

AMPHIBIOUS GOBIES.

We need not go to foreign lands to find a goby that does not object to a trip overland, as on our Southern shores, Texas and Mexico, are found several species that are often seen in shallow pools left by the tide. One especially, the *Gobius soporator*, will of its own accord crawl over the moss and weed, thus passing from pool to pool. During a recent government expedition to the Texas and Central American coast, numbers of these little fishes were captured by the naturalists in charge, and placed in buckets some distance from the water, preparatory to taking aboard the steamer. The gobies, however, had different views; and without loss of time, one and all began to clamber out of the pails, and were soon on their way overland to their natural element. When recaptured they immediately began their attempts at escape, and seemed but little inconvenienced by the time spent out of the water. (*Plate VI.*)

A fish found in Ceylon is often seen out of water, and when the pool in which they live becomes shallow, the fishes burrow in the mud, working their way downward sometimes to a depth of four feet. If the drought penetrates to them there, they wriggle to the surface again, and in a body move into the woods in search of water, and by some peculiar instinct they generally travel in the right direction. By the aid of the grass they are enabled to keep an upright position, slowly moving along by means of the pectoral fins, that in this family are very strong and long. Some jump and use the tail, but the general motion is by a backward and forward movement of the fins.

THE CLIMBING PERCH.

That certain fishes were inclined to live on shore was well known to the ancients, and Theophrastus is supposed to be the author of a work entitled " Fishes that Lived on Dry Land." He says that in India certain little fishes resembling the mullet leave the rivers for a time and return to them again. Although a commentary on this treatise was published in 1665 at Naples, by Severinus, it was not properly understood until 1797, when M. Daldorf communicated to the Linnæan Society his observations on the tree-climber (*Anabas scandens*), one of which he had himself captured as it was ascending a palm tree that grew near a pond. The object of the fish in making this ascent is said to have been to reach a small reservoir of rain water collected in the axils of the leaves, and full of insects. This faculty of climbing has been vouched for by some observers, and denied by several.

Dr. Cantor says that in the Malay countries the Anabas is eaten by the poorer classes, who do not attribute to it either the medicinal qualities or the climbing propensities for which it is celebrated. It can live long, however, out of water, and is frequently sent in a dry vessel from the marshes of Jazar to Calcutta, a journey of several days, which it survives. (*Plate VII.*)

By what special provision of nature these and other fishes are enabled to adapt themselves to an amphibious existence has for a long time been little understood. We see ordinary fishes breathing by their gills, absorbing oxygen from the current that bathes them, then passing out behind the operculum through the gill-opening ; but in various forms we find certain modifications.

In the families *Luciocephalidae* and *Labyrinthici* the gills or breathing-organs are modified, and have cavities with peculiar folds, which extend far into the head ; while in the *Ophiocephalidae*, the species have a simple cavity with feebly developed folds.

The generally accepted theory of their power of living upon land has been explained by assuming the hypothesis that the labyrinthine cavity was used to store water to be used in their wanderings upon land. Recent researches, however, have shown the fallacy of such a belief, and the accessory gill-cavities are thought never to contain water, but are air-cavities that take the place of true lungs ; a fact that shows the Boleopthalmus and others to be truer amphibians than even the frogs.

The curious cat-fishes, Doras, and Callichthys, are noted overland travelers. In the dry seasons the streams in which the latter is found run low, when a remarkable scene is enacted ; the entire body of fish start overland, a compulsory migration, but with unerring instinct they head for distant water. At times the column that is struggling through the grass, now erect, now on their sides, comes to a halt, and some of the fish burrow, as if with the intent of finding water below the surface. Birds and other animals prey upon them, but eventually, they reach water, not having been affected by their stay on dry land. (*Plate VIII.*)

Another cat-fish, found in South American streams, seems also at home out of water. Voyagers have frequently observed them floating down the streams upon submerged logs, upon which they had crawled after the manner of frogs. In appearance they are extremely striking, the head seemingly ornamented with an array of writhing snakes, in reality the whiskers or feelers of the Tangsa.

During the dry season of Africa and South America, the streams in which the *Dipnoi*, or lung-fishes live are often dry, and at the first approach of what would seem a dire calamity the fishes retreat to the bottom, forming a cell in which they pass months in a state resembling hibernation. The cases or nests have been sent to Europe in trunks and the fishes then soaked out. (*Plate IX.*)

Though not amphibious in their habits, there are a number of fishes that obtain their food out of the water ; such is the archer fish, *Toxotes*, whose extended lower jaw seems perfectly adapted for such work, and, swimming along shore, if the fish observes an insect it rises instantly and ejects a drop of water with unerring aim three or four feet, bringing the insect to the surface, where it is snapped up by the finny marksman.

The long-beaked Chaetodon obtains much of its food in a similar manner, the long bill serving to guide the drop, thus being a veritable blow-gun. These fishes are often kept in aquariums in the East, and so tamed that they will shoot insects held in the hands of their owners over the water.

The Anableps, a South American fish, obtains much of its food upon the surface of the water, and to further this purpose, has eyes that are so divided that the fish possesses apparently four, owing to the cornea and iris being divided by transverse bands, so that the two pupils are observed upon each side, while the other parts of the eye are single. From this peculiarity they are known on the rivers of Guiana as four-eyed fish ; but the modification is undoubtedly to enable them to secure prey upon the surface, their movements being much like those of a frog leaping along upon the surface partly out of the water, so that they would be taken by the casual observer for these animals.

CHAPTER IV.

THE METEORS OF THE SEA.

ONE of the most fascinating themes connected with our knowledge of the ocean world is that bearing upon its illumination. Until quite recently it has been thought that animal life could not exist in the abyssal depths of the sea, but recent American and European expeditions have shown that this is not the case, and this obscure region, that was supposed to be the most desolate spot on the habitable globe, is found teeming with life, the numerous forms adapted by peculiar modifications of structure to their life in a world over which roll perhaps four or five miles of water, where the temperature is just above freezing, and where the pressure amounts to two or three tons to the square inch of surface.

When considering the conditions of such a life, we wonder how the inhabitants of this dark region see to move about, and whether nature has made any provision for their wants. In answer we need only visit the seashore, and there we shall find that the old ocean instead of being wrapped in darkness, has its moons, suns, and stars of living lights that illuminate the greater or less depths with their splendors. Many of them are the common objects of the seaside, and in our wanderings on summer nights may always be observed, especially on rocky coasts where the beauties of the phosphorescent jelly-fishes are to be seen. The dark surges come thundering on in tidal measure, laden with the secrets of the sea; the crest seemingly ignites,

40

combing, gleaming on the rise far down the line, and then with sullen roar is hurled a mass of living light upon the sands, trickling back in rills of molten gold only to storm the breach again and again.

On the New England coast these displays of phosphorescent phenomena are particularly noticeable. When

> "The day is done, and the darkness
> Falls from the wings of night,"

the phantoms of this world of light spring into existence, changing the bosom of the ocean to a scene of weird revelry. Every drop of water seems a gleam of light; and the brown kelps and sea-weeds that hang upon the rocks drip with liquid fire. Ahead of our boat as we slowly scull along, waves of light appear; while beneath, moons and stars move here and there, revolving and rising in graceful curves with gentle undulation. Now swift flashes, coming from the gloom beyond, dart across the field, leaving a brilliant nebulous train behind.

The scene as the waves break upon the rocks is one of dazzling splendor. At Spouting Horn, Nahant, the water forced through a natural crevice in the overhanging crag, is thrown high in the air; for a moment it hangs suspended—a luminous mist, then settles upon the grim battlements, bathing them in a warm lambent light that winds its way in gleaming rivulets to the sea. We dip our scoop-net into the water; the wish of old King Midas seems here fulfilled. The meshes become a sheeny web of golden fabric, and in the catch are myriads of gleaming, living creatures, the veritable lamps of the sea.

They are *Medusae*, jelly-fishes, if you will; unsightly objects when stranded upon the shore, but at night possessed of a loveliness peculiarly their own. Large forms of Aurelia and Cyanea are striking objects as they glide along, surrounded by a halo of golden greenish light.

The Cyanea is a giant of its kind, a fiery comet moving in and out among the lesser constellations. One of these huge jellies, observed near Nantucket, from the mast of a vessel, was seen swimming lazily along, its disk surrounded by a luminous halo, fifteen feet in diameter, while the train of gleaming tentacles stretched away two hundred feet or more.

One of the most interesting exhibitions of the light of a jelly-fish was witnessed by Mr. Telfair in 1840, near Bombay. The natives had reported at various times that a gigantic flaming monster had been seen in the sea, and some said that they had observed it in the sky many years before—evidently meaning a comet. Finally Mr. Telfair himself saw the monster that proved to be a jelly-fish of enormous dimensions. Its tentacles at night seemed a fiery train over three hundred feet in length, presenting a dazzling spectacle to those who rowed over it ; each tentacle appeared like a red-hot wire, gleaming with a brilliant light, while the body resembled an enormous incandescent sphere throwing out a light for many feet about it. The jelly finally ran ashore upon the beach, or was washed in, where many of the natives watched it, hoping to obtain its skeleton, which it is needless to say they failed to find. For several days after it stranded it emitted so strong a light at night that it was visible for a great distance, and illuminated the forms and faces of those who stood about with great brilliancy. It was estimated to weigh, including all the tentacles, two tons, and was the largest invertebrate animal ever seen. In the daytime the great semi-transparent disk of the Cyanea with its flexible lobed margin is a dark reddish-brown color, while the tentacles, bristling with lasso-cells and spiral darts, are yellow, purple, brown or pink.

While the Cyaneas tint the sea with a greenish light, the little *Dysmorphosa,* that at times appears in vast numbers around rocky points, illumines it with a deep aurelian hue. On successive nights we may find as many different varieties of these jelly light-givers changing the water to white and yellow tints. The shapely *Zygodactylae* produce a strange effect as we see them in the depth below wandering about—the *ignus fatui* of the sea. Here the delicate *Idylia* gleams with ever changing hues; *Pleurobrachia* rise in graceful curves, their fringed trains glistening with red, green, yellow and purple rays; the golden *Melicerta* flit by with erratic course, while the resplendent forms of *Coryne, Tima, Clytia, Eucope,* and a host of others add to the glories of the scene.

The *Pleurobrachia* and its kindred, from the peculiar character of their locomotive appendages, are among the most beautiful of all marine light-givers. The *Beröes* are perhaps the most familiar; assuming many shapes, sometimes spherical, oval and oblong, they move through the water, clear as crystal, by means of their lace-like hyaline fins, which, as they wave to and fro in rhythmic measure, decompose the rays of light from other forms, and glitter with hues of vivid iridescence.

So beautiful are these creatures that they have been immortalized in verse by Drummond in the following :

> "Shaped as bard's fancy shapes the small balloon,
> To bear some sylph or fay beyond the moon.
> From all her bands see lurid fringes play,
> That glance and sparkle in the solar ray
> With iridescent hues. Now round and round
> She whirls and twirls; now mounts, then sinks profound."

So vast are the numbers of these and other light-giving seajellies in the Northern seas, that the olive-green tint of the waters is due to them in the day time. Mr. Scoresby finding sixty-five of them in a cubic inch of water, summed up the amusing calculation that if eighty thousand persons had commenced at the beginning of the world to count, they would barely at the present time have completed the enumeration of individuals of a single species found in a cubical mile.

These beautiful phosphorescent jellies can be observed as we drift along, by using a small glass cylinder. Keeping the finger pressed upon the top, and lowering the open end near the little creature, then removing the finger, it will be drawn into the improvised aquarium. If the night is very dark, the play of light about its delicate form will be found a rare study.

Darwin refers to the beauties of the phosphorescent jellies observed on one of his collecting tours. He says : "While sailing a little south of the Plata on one very dark night, the sea presented a wonderful and most beautiful spectacle. There was a fresh breeze, and every part of the surface, which during the day is seen as foam, now glowed with a pale light. The vessel drove before her bows two billows like liquid phosphorus, and in her wake she was followed by a milky train. As far as the eye reached the crest of every wave was bright, and the sky above the horizon, from the reflected glare of these livid flames, was not so utterly obscure as over the vault of the heavens. . . . Having used the net during one night, I allowed it to become partially dry, and having occasion twelve hours afterward to employ it again, I found the whole surface sparkling as brightly as when first taken out of the water. It does not appear probable, in this case, that the particles could have remained so long alive. On one occasion, having kept a jelly-fish of the genus *Dianæa* till it was dead, the water in which it was placed became luminous."

To Spallanzani is due the credit of first calling attention to the phosphorescence of jelly-fishes. He made some interesting experiments with the jelly *Aurelia phosphorea*, and came to the conclusion that the light-emitting power lay in the arms, tentacles, and muscular zone of the body, and cavity of the stomach; the rest of the animal showing no luminosity. The light seemed to be a viscous liquid, a secretion that oozed to the surface. One aurelia that he squeezed in twenty-seven ounces of milk made the whole so luminous that a letter was read by the light three feet distant, being the first practical result of the discovery of animal phosphorescence. Humboldt experimented with the same animal, and, having placed it upon a tin plate, observed that whenever he struck it with another metal, the slightest vibration of the tin rendered the animal completely luminous. He also observed that it emitted a greater light when in a galvanic circuit.

Of the *Infusoria*, the giant monad *Noctiluca* presents the most gorgeous spectacle. It can hardly be seen with the naked eye, and resembles a currant; a curved filament, its locomotive organ, resembling the stem; beneath the outer envelope is the gelatinous layer containing numerous granules, that seem to be the light-giving organs. A gobletful of these minute creatures produces sufficient light to read by at a distance of two feet, the glass appearing fairly ablaze with light, while a sensitive thermometer placed among them shows not the slightest elevation of temperature.

Humboldt bathed among the *Noctilucae* of the Pacific, and tells us that his body was luminous for hours after, and even the sands upon which they were left at low tide appeared like grains of gold. The captain of an American vessel traversed a zone of these animals in the Indian ocean nearly thirty miles in extent.

It was a perfect night, yet the light emitted by these myriads of fire-bodies eclipsed the brightest stars; the milky way was but dimly seen, and as far as the eye could reach the water presented the appearance of a vast gleaming sea of molten metal of purest white. The sails, masts and rigging cast weird shadows all about; flames sprang from the bow as the ship surged along, and great waves of living light spread out ahead—a fascinating and appalling sight.

I have observed the beauties of these forms in the tropics, but as well on the Maine coast, and in a rocky cove near York the display was often remarkable. Returning from a day's fishing late one night, I noticed a number of brilliant lights, apparently on the rocks. Impressed with the belief that our friends had lighted lanterns and were fishing, we rowed in that direction, only to find that the lights were bunches of pendent sea-weed, glowing as if at a white heat, the light produced from myriads of *Noctilucae* and other luminous animals left on the weed by the ebbing tide. So brilliant was the phosphorescence on this evening, that by merely splashing the water with my hand, the sea seemed to ignite and blaze in living flames.

The *Noctilucae* alone do not always produce these wonderful effects; vast shoals of diatoms illuminate the Southern seas, those known as *Pyrocistis pseudonoctiluca* and *P. fusiformis* being the most brilliant. They are generally confused with the former animals, resembling them somewhat in general appearance, but have an extremely thin coating of silica, and differ from them in many ways. The magnificent spectacle of a water-spout rising among these forms has been seen, resembling a gigantic pillar of fire, followed by a cloud of seemingly luminous foam and spray. Among the peculiar bottom light-givers, the star-fishes *Asterias* and *Ophiura* are worthy of notice, often being seen gleaming with a pale phantom light. Of them the late Sir Wyville Thompson says: "We were steaming slowly back toward the coast of Ireland; and on July 26th we dredged in depths varying from five hundred and fifty-seven to five hundred and eighty-four fathoms in ooze, with a mixture of sand and dead shells. In these dredgings we got one or two very interesting alcyonarian zoophytes and several ophmirideans. Many of the animals were most brilliantly phosphorescent, and we were afterward even more struck by this phenomenon in our Northern waters. In some places nearly everything brought up seemed to emit light, and the mud itself was perfectly full of luminous specks. The alcyonarians, the brittle stars, and some annelids were the most brilliant."

On the trip to Shetland from Stornaway they were fortunate in observing another exhibition of the phosphorescent phenomena. He says: "Among star-fishes, *Ophiacantha spinulosa* was one of the prevailing forms, and we were greatly struck with the brilliancy of its phosphorescence. Some of these hauls were taken late in the evening, and the tangles were sprinkled over with stars of the most brilliant uranium green; little stars, for the phosphorescent light was much more vivid in the younger and smaller individuals. The light was not constant, nor continuous all over the star, but sometimes it struck out a line of fire all round the disk, flashing, or one might rather say glowing up to the centre; then that would fade, and a defined patch break out in the middle of an arm and travel slowly out to the point, or the whole five rays would light up at the ends and spread the fire inward."

The sea-pens are remarkable for their phosphorescence, *Renilla reniformis*, according to Agassiz, emitting a golden-green light of most wonderful softness.

The *Pennatula phosphorea*, as its name implies, is a wondrous light-giver; the luminous halo assuming a rich purple tint as they move along by the regular pulsation of their fringed arms that, when fully expanded, resemble the feathers of a quill pen. One of the most beautiful of these forms is the *Virgularia mirabilis*. "Two series of half-moon shaped wings, obliquely horizontal, are placed symmetrically round an upright axis. They embrace the stem somewhat in the manner termed *petiolate* by botanists, clasping it alternately; or, shall we say, like two broad ribbons rolled round a stem in an inverse direction, in such a manner as to produce the effect of two opposing flights of stairs. These wings are waving, vandyked, and fringed on their outer edge, and of a brilliant yellow; the dentature of the fringe being the lodging of their pretty little polyps, which display occasionally their gaping mouths and expanded frills. The polyps are white and semi-transparent. When they display their rays, the margin of each wing presents an edging of silvery stars." A huge Arctic form, the *Umbellularia*, also a light-giver, grows to the height of four or five feet. The sea-pen *Virgularia* is often left bare at low tide on the Patagonian coast and at night the shoal appears dotted with myriads of torches, resembling the lights of an immense army.

Many of the sea-pens emit a lambent white light, while others give out gleams of many tints. The scientists of the "Porcupine" observed a most brilliant display near the coast of Scotland. Sir Wyville Thompson writes: "Coming down the sound of Skye from Loch Torridon, on our return, we dredged in about one hundred fathoms, and the dredge came up tangled with the long pink stems of the singular sea-pen *Pavonaria quadrangularis*. Every one of these was embraced and strangled by the twining arms of *Asteronyx loreni*, and the round,

THE WALKING LEAF—(PHYLLIUM SICCIFOLIUM).

PLATE XXI.

PLATE VI.

AMERICAN GOBIES CRAWLING ON THE SHORE.

soft bodies of the star-fishes hung from them like plump, ripe fruit. The *Pavonariæ* were resplendent with a pale, lilac phosphorescence, like the flame of cyanogen gas; not scintillating, but almost constant, sometimes flashing out at one point more brightly, and then dying gradually into comparative dimness, but always sufficiently bright to make every portion of a stem caught in the tangles or sticking to the ropes distinctly visible. From the number of specimens of *Pavonariæ* brought up at one haul, we had evidently passed over a forest of them. The stems were a metre long, fringed with hundreds of polyps."

When the ship "Venus" was lying off Simonstown, one of her boats passed over a forest of sea-pens in shoal water, that gave such a light that the crew could read by it; while where the ship lay at anchor other forms of phosphorescent animals illuminated the ports so that the men laid in them, and read by the wondrous light on the darkest night.

4

CHAPTER V.

PARENTAL CARE AMONG ANIMALS.

In their care of offspring the lower animals perhaps approach nearer to the human standard than in any other respect. Even in the lowest forms, where action seems subservient to instinct alone, we find acts of maternal or paternal devotion curiously similar in their performance to those of man. What care exceeds that of the common spider, that, when alarmed for the safety of her progeny, grasps the silken nursery in her mandibles and rushes away, and when brought to bay fiercely contends against superior forces to the last, often being torn limb from limb before relinquishing her hold?

The young of a small black spider common in the New England States are protected in a curious manner. The mother carries them about on her back, the clinging mass often completely covering her; when alarmed and closely pursued, and it becomes necessary to render the flight less conspicuous, each of the young springs from her back in different directions, first having attached to her a silken cable, by which apron-string they find their way back.

A spider in the East envelops her eggs in an oval balloon, to which a silken rope is attached and made fast to a leaf or twig, and floats securely in the air, by its motion defying the most active of its enemies. In almost every family of insects the same care is noted. One of the centipedes actually sits upon its eggs after the manner of a hen; rolling them over and

50

over with its many feet to remove any fungus that might adhere
and prove detrimental to them later on.

The mole-cricket, *Gryllotalpa vulgaris*, has earned a reputa-
tion as a good-hearted insect, showing great solicitude for its
young. These little iron-jawed creatures are very common in
the South, and perhaps it will be of interest to insect collectors
to know that they have obtained so firm a footing in some of
the extreme outer keys of the Florida reef that it is next to
impossible to raise anything. Our garden was raided by them
day and night, the loose coraline sand affording a medium
through which they moved with great rapidity, the upturned
ridges resembling those of the mole. On account of the nature
of the soil, being entirely of ground coral, and the plates of
lime-secreting algæ, mixed with broken shells, I never could
find that they formed a nest in this locality, and as during
heavy gales the tide sometimes rose up through the key until it
was a foot deep in front of our house, the nests would hardly
have lasted. But in various localities on the mainland, where
they have been observed, their care and solicitude for the young
is indeed marked; no animal, in fact, outdoing them in this
respect, and so, although I fought them for five years or more,
I will not refrain from giving them proper recognition. When
the eggs are about to be deposited, and this refers to localities
within the frost-line, the female erects or forms a cell of earth,
about as large as a hen's egg, that she packs together for the
purpose, in the centre of which she places the coming young.
The cell becomes hardened, and is in itself a sufficient protection, but the cricket does not depend upon this alone, as about it
is constructed a perfect moat, while innumerable avenues and
blind leads are built to mislead the various carnivorous beetles
that are ever on the lookout for such luxuries as young crickets

or eggs. It is quite evident that the mother displays something akin to thought ; at least, she appreciates changes of weather, as does a cold wave come suddenly upon the locality in which she is, she immediately removes the family to a deeper level, so that the growth of the immature young may not be retarded. As soon as the cold wave passes they are again brought nearer the surface where the genial rays of the sun reach them, at least in effect. Almost equally solicitous for the welfare of their young are the ants. The leaf-cutting species of South America are often seen in long columns bearing pieces of leaves to their vast subterranean nests, that, when arranged as thatching or otherwise, become overgrown with fungi particularly adapted as food for the young, and eaten by them with great avidity. At the slightest warning, the young ants or eggs are seized by the workers and conveyed to a place of safety, and cared for with all the tenderness displayed by parents of a higher order of intelligence. Some of the large African workers will submit to all kinds of torture before releasing an egg. Their legs may be severed, body, and finally the head, which will retain its hold upon the unconscious offspring for hours after.

The white ant (*Termitidae*), a neuropterous insect, shows great solicitude for the young, and remarkable preparations are made for their proper care. To this end they erect structures that in Africa have been observed fifteen feet in height, the entire hill covering an area of twenty-five square feet. In the centre of the pile is a cell resembling the section of half an egg, originally about an inch in length, but as the queen grows it is enlarged to meet her requirements, and they are not a few. In mature life she is a thousand times heavier than any of the servitors that wait upon her. The royal chamber, as it is called, is built of clay, so that it becomes perfectly hard and durable. A number of doors are left just large enough to admit the laborers and soldiers, but the queen herself is a prisoner. About this chamber are various rooms much smaller, generally nicely arched, but of different shapes. These are the quarters of the soldiers and laborers, where they sleep, rest, and stand guard upon the queen. These rooms are all connected, and branch off into a labyrinth of paths and lanes that lead to store-houses where the food supply is cared for. Near the royal chamber are found quite a different kind of apartments. They are very small, but built in large clay chambers, and are formed of woody fibre joined by some gummy secretion. These are the nurseries, where the young and eggs are carefully tended, and, curiously enough, the walls of these minute rooms are generally found overgrown with a delicate mould, dotted here and there with little white globules about the size of a pin's head. The latter are microscopic plants, that under the glass look like mushrooms, and it is supposed that in some way their growth is encouraged here to furnish a tender and succulent food supply for the young when they first appear. To give an idea how fast they do appear, it may be said with safety that the queen deposits at least eighty thousand eggs a day, that are immediately carried away by the attendants and placed in the nurseries that are continually being built for them, and where they are watched and cared for by vast numbers of workers.

The female scorpion bears her young about upon her back, and, according to some authorities, they repay this care by devouring the mother

The habit of carrying young upon the back as a protection, is seen in a large and varied class of animals. We have noted the spider and its young, and in strange analogy is the opossum, the common marsupial of the South. At first the young are retained in the pouch. When not alarmed they appear in various strange positions on the mother's back, their smooth, prehensile tails coiled about that of the parent that is, perhaps, bent over her back for the purpose; the tail seemingly serving as a fifth limb, intuitively clasping branch or bough.

The great ant-eater, a thoroughly clumsy creature, walking upon the sides of its clawed fore-feet, transports its young on its back, which performance I have been fortunate in witnessing. The young ant-eater clings to the rough fur, throwing its tail forward over its head, while over all comes the bush-like canopy of the mother, forming effective concealment to her long-nosed offspring.

The huge hippopotamus has been observed by many travelers, drifting down the sluggish streams of the East, bearing upon its broad, platform-like back, a pink, shapeless young, and often when in deep water the gigantic infant appears to be floating along lightly upon the very surface of the water. In a similar way young alligators are borne about, often thus becoming exposed while the parent is hidden below.

Among the tree-toads are several that carry about their young in a like manner, especially those found in the islands of Guadaloupe and Martinique. (*Plate X.*) In the latter, owing to a lack of swamps and water suitable for the proper development of the young, they are carried about by the parent, clinging to its back by some peculiar secretion. In the nototrema there is a sac in the back that contains the young; but perhaps the most interesting case is that of the Surinam toad. In the breeding season the female deposits her eggs in some secluded place, and instead of leaving them, after the manner of many mothers, she remains in the same spot, until the male with his broad,

web-like feet lifts them upon her back, where they are retained
by some glutinous secretion. Now a curious change takes
place. The eggs gradually disappear, seemingly being absorbed,
hexagonal shaped cells forming around them. At this period
the toad enters the pond and conceals herself in the mud ; the
skin that supports the eggs now becomes inflamed, and the
cells finally become covered with a thick membrane, the eggs
entirely disappearing, and the back of the animal resembling
a piece of honeycomb more than anything else, the cells
being about large enough to admit a large horse-bean. When
the young have sufficiently grown, the mother leaves the pond
and crawls upon the shore, when a strange scene is enacted.
The young toads are seen leaving the cells in all positions—
some head first, with legs and arms protruding ; others clinging
to her back as if loath to leave, while many more plunge off
into the mud and water, becoming food for birds and fishes.
In eighty days all signs of this curious performance have disappeared, the cells becoming absorbed, and only reappearing on
the return of the breeding season again.

The Aspredo, a South American catfish, carries its eggs about
with it. During the breeding season, and after the eggs have
been deposited, the fish passes over them, the eggs becoming
attached to the ventral surface and fins in great numbers.
Horny, stalked peduncles connect most of them, so that the eggs
dangle like pendants, all traces of the curious nurseries disappearing after the young are hatched.

Another catfish, found at Panama, has a sac-like fold in the
skin of the abdomen, in which the young are carried. In the
sea-horse this is even more striking. As soon as the eggs are
deposited, the male, who possesses the pouch, in some way
receives them into it, and the young are nurtured by its fatty

lining, often as many as a thousand young colts, measuring about five lines in length, being so cared for. When, in the judgment of the parent, they are sufficiently grown to swim about in safety, the sac is pressed against a stone or shell, and the young brood are forced out of their nest, presenting a curious spectacle as they move along like a cloud by the rapid, vibratory movement of their minute dorsal fins.

During his journey in Brazil, Professor Agassiz discovered a fish allied to the cat-fishes, that not only carried its eggs in its mouth and gill-folds, but the living fish were found there in great numbers. This is equally true of an East Indian species of *Arius*, and of a large and varied class of animals.

A number of the echinoderms, discovered by the "Challenger" expedition, were provided with a nursery similar to that of the sea-horse, called a marsupium. In some the spot was covered by thick plates, that were gradually forced up, forming effective protection to the young; in others the long spines were directed over the spot, embracing and imprisoning the bristling and spinous progeny.

The curious crustacean, *Arcturus*, common in Arctic seas, bears its young upon its long claws that are raised above and before its head.

The domestic cat is a more familiar example of animals that bear their young in their mouths, the mother often performing wonderful feats in the way of transporting her family, carrying the weakest between her teeth, while encouraging the others to follow. An interesting case came under my observation, showing their persistence.

A dog arrived at a farm-house unfortunately at a time when the old family cat was engaged in the earliest maternal duties. The next day she and her litter of five were missing, and word

PLATE VII.

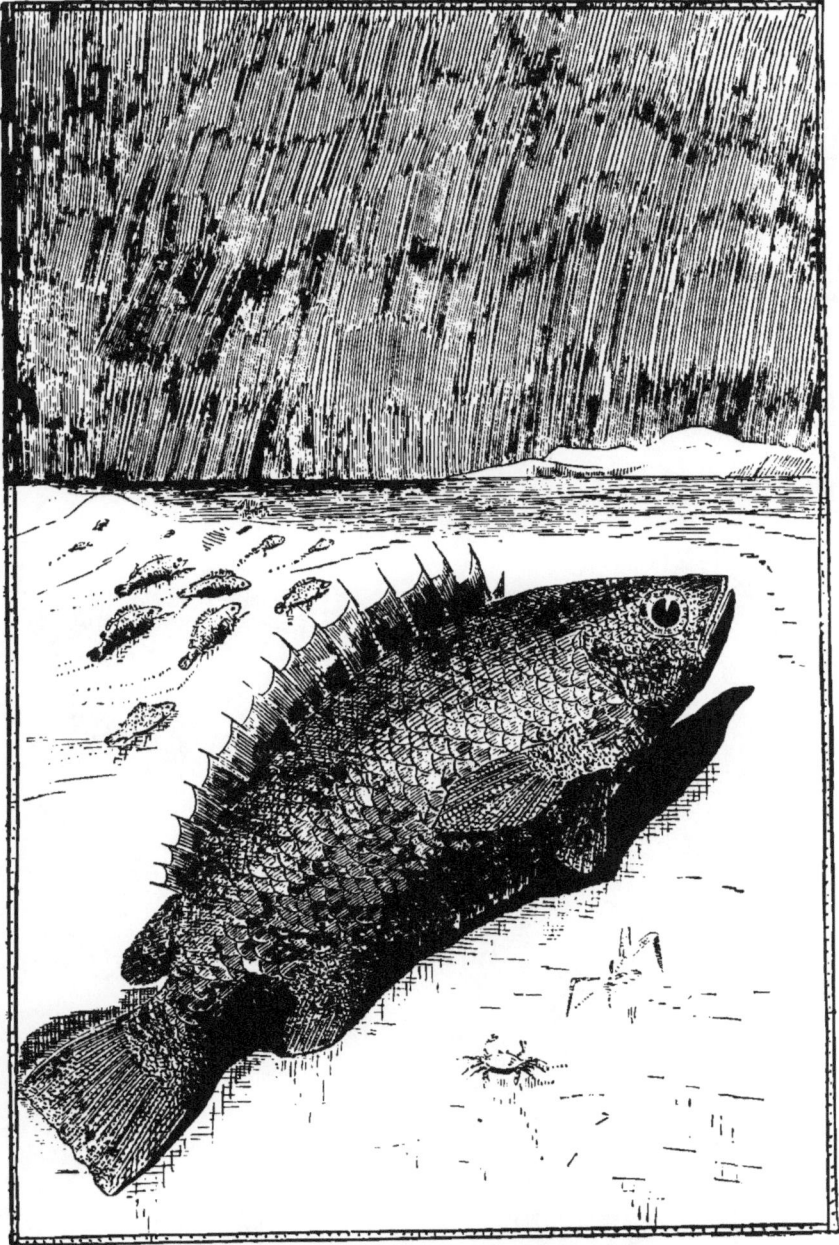

CLIMBING PERCH MIGRATING OVER LAND.

was brought from the nearest neighbor, a mile away, that the old cat was there. She and her family were brought back and watched, and almost immediately the mother seized one of the kittens in her mouth, and, head high in air, started off through the fields to the friendly neighbors that did not keep a dog. In the course of the day the entire family were safely removed a second time, the cat having traveled during the transportation ten miles. Suffice it to say she was allowed to remain until the departure of the possible enemy.

Hunters in India have observed the tiger, generally so ready to stand her ground, slink away with something in her mouth, at first supposed to be prey, but closer examination showing it to be her young, and in all the cat tribe this same trait is seen.

Among birds the most painstaking endeavors are seen in the erection of their nests, that are built in a variety of ways, to afford all possible protection to the young. Many of the humming-birds' nests are covered with moss taken from the tree upon which they are built, and so skilfully adjusted that they mimic the tree, and can scarcely be distinguished from it. Others are fastened upon leaves that, constantly moving, afford protection.

Certain birds related to the raven cover their nests with a *chevaux-de-frise* of briers that protects the young from predatory animals. In Africa others that are preyed upon by snakes build long pendent nests over the water, the opening being low. One of the grebes builds a floating nest that rises and falls with the tide, and can be paddled away by the mother. But perhaps the most astonishing instance of maternal care among birds is that observed in the woodcock. The mother bird has been seen by a number of sportsmen, when closely followed by them, to rise with a single young between her feet and fly heavily away.

The common snipe displays almost equal intelligence. When her nest is approached she feigns lameness, and hops off clumsily in an opposite direction, and when the nest is far behind, she assumes her natural gait, takes wing and flies off to regain it by a roundabout way.

Many of the penguins have a pouch in which they carry their single egg until hatched, thus bearing their nest about with them. At this time their motion is a hop, the feet being kept together to hold the egg in place, but when the young is hatched they walk as do other birds. The albatross, although it builds a high nest, also conceals its egg in a fold in the skin, so that it is difficult to ascertain whether they are sitting even when lifted from the ground.

A number of the stormy petrels rear their young underground, after the fashion of the burrowing owl. Here, however, the young birds have a natural outlet, but in the Celebes bird *Maleo* the eggs are buried several feet in the sand along the beach, exactly as are turtle eggs, the bird showing great cunning in destroying her tracks to the eggs, a peculiarity that I have noticed in the green turtle. The eggs are finally hatched by the heat of the sun, the young birds digging their way up to the surface, and, strange to say, they are enabled to fly immediately, a necessary provision, as the maternal duties end with the burial of the eggs. These strange birds are allied to the mound-building *Megapodius* that has somewhat similar habits.

Among the seals and other marine mammals the young are often held or supported by the flipper, so that when standing upright in the water, embracing their curious young, they bear much resemblance to the typical mermaid. To such occurrences are we indebted for many of the marvellous tales related by the mariners of the olden time, who, no doubt, believed that such evidences of affection were impossible among common animals.

Even among the shells wonderful provision is seen for the protection of the young. In the argonaut the eggs are fastened to the interior of the pearly home. The violet snail forms a raft to which its eggs are attached, the entire family floating along in company. Other shells carry about their young on the capacious foot. The natica moulds its eggs in a collar of sand, while the great land snails of South America, as the bulimus, form regular nests of leaves in which their great bird-like eggs are deposited. In some cases they are laid in rows upon a single leaf, the latter being rolled up over them. In the crabs the eggs are generally attached to the abdominal limbs, while others deposit them in the sand, or carry the young in the immature state, clinging to their back.

The wonderful foresight of insects in depositing their eggs in places favorable to the young when hatched, is called instinct, but many observers see thought and intelligence in the action. A wonderful instinct is that which causes certain insects whose young depend upon the hives of honey-bees in which to pass a period of their existence, to deposit their eggs on certain flowers, so that the young larvæ may clasp the visiting bee, and thus be transported to its storehouse. Many of the ichneumon flies penetrate the bark of trees, beneath which a grub is safely ensconced. The egg is placed in it, the young feeding upon the victim later on; and thus millions of caterpillars and grubs become living nurseries and future food for the young of many species.

Some of the wasps capture other insects, and after paralyzing them deposit in the body an egg, then burying it, the insect remaining alive but motionless until the birth of the young wasp, when it is slowly devoured. Other insects, knowing that their young require dead wood, deposit their eggs in a limb, and carefully girdle the branch below them, thus preventing the flow of sap, and by the time the eggs hatch they are in a dead limb, through which the larvæ work with ease.

The gall insects are provided with some secretion that is deposited with the egg in a tender branch, causing an abnormal growth about it, forming perfect protection for the coming larvæ, and more especially the exact food that it requires.

In all these cases, and many more, instinct seems to play an important part; but who shall say that the animals do not possess thoughts and desires differing from our own perhaps only in degree.

CHAPTER VI.

AN OCEAN SWORDSMAN.

"It wan't so fur from here thet I run afoul o' my first so'd-fish," said Captain Sam, leaning on the wheel with one booted leg over a leeward spoke, while a smile at some long-forgotten memory radiated all over his bronzed countenance. "I was a yonker then," he continued, "and, ef I dew say it, was dretful green—dretful—a regular high-tide clam-digger, always a-lookin' fur the wind from lew'ard, and the like."

"Starboard," came from aloft in lusty tones, and Captain Sam sprang into activity, and with a jerk at the wheel sang out, "Where away?" "Tew pints ter lee'ard," was the answer; and quickly the little vessel fell away, gathering fresh energy under the slackened sheet.

We were off Boon Island, Maine. Our craft, the Thumb-screw, as Captain Sam said, "rated A No. 1 in York county—three decks and no bottom, no odds wanted, and none taken."

The Thumbscrew was a sword-fisherman, long, low and rakish, fast, and wet in rough weather, differing from other vessels of her kind only in the iron stanchion that ornamented the tip end of the bow-sprit, in which the harpooner found support and security when wielding his lily-iron. Captain Sam, who whiled away the long hours with an incessant flow of original volubility, and told us privately, "when I ain't a-talkin' you'll know I'm sick," was skipper, and four men, who hailed from down East, constituted the port and starboard watches, first, second, and third mates, cook and crew.

Another hail coming from the foretop, one of the crew ran nimbly out upon the bow-sprit, and, leaning against the semi-circular iron band that tipped the stanchion, unlashed the long harpoon, which, from the character of the prong, is called the lily-iron. The rest of the hands saw that the rope attached to the pole was not fouled with the downhaul, that the coil in the tub was in shape, and that the gayly-painted keg at the end of the rope was ready for a bath. By the time these minor details had been attended to, the sharp dorsal fin of a huge fish could be seen from the deck, cutting the sea, and rushing about in an erratic manner.

"He's takin' his last meal, ef the wind holds," said Captain Sam, with one eye on the sword-fish and the other instinctively on the luff of the main-sail. "Curious fish. I've hed 'em give me a chase fur hours right in sight, and couldn't fetch 'em over the bow or alongside tew save me. Fall away, so would they. If we're tew lew'ard, and luffed or tried tew eat up on 'em fish would haul tew wind'ard too."

The sword-fish was now moving along about a hundred yards to windward, and by going aloft a bird's-eye view was obtained of its movements. It was nearly ten feet long, and had charged a school of mackerel, which in the summer months throng the Gulf of Maine. Now the lithe fish would dash at the flying school, that in turning seemed a galaxy of silvery stars gleaming in a watery sky. As they massed in terror, they seemed to give out flashes of light, moving in circles or in graceful curves, suddenly disappearing, again leaping into the air in wild affright, falling back to join the panic-stricken throng that rushed on, the ocean before them, but safety nowhere. Once in the fleeing mass, the swordsman made savage cuts to right and left, up and down, now leaping half out of the water in tierce and carte, thrusting, striking, and plunging in savage enjoyment, while a shower of sinking silvery bodies told of the carnage done. Then, lightly turning, and showing the white ventral surface, the great fish deftly picked up the wounded victims and severed parts, occasionally striking them again with its sword, as if to impart some of the excitement of the chase to the more prosaic after-play of eating. Not all the spoils were collected; the living game had greater attractions, and, rising, the great dorsal fin was soon cutting the water in pursuit, startling the white-winged gulls that in eager expectation were following—the ghouls of this marine battle-field.

Gradually the skipper had been keeping away and gaining on the fish, and now, with a mighty turn of the wheel, the little vessel fell away with a rush, the jib being hauled hard to windward, and the fish was across the bow. The long pole quivered in mid-air, the filed barb of the steel lily glistened, then with a splashing thud was buried in the body of the fish.

"Stand clear the line!" shouted the mate. Captain Sam was winding down the wheel, and the schooner fairly groaned with the suddenness with which she was brought up into the wind, the line shrieking and singing as it tore from the tub in its flying race after the fish.

"Thet was a master hit," said Captain Sam. "Right afore the side fin; sent in tew stay." "I calculate 'twas," replied the harpooner, who had lashed the pole and now took in hand the barrel and held it aloft.

"All ready!" And with a jerk the last coil leaped from the tub, and the keg was thrown into the sea, to follow and eventually wear out the gamey fish.

The fore and main gaff topsails were now clapped on to the Thumbscrew, and with the white spot moving ahead for bearing, she plunged on in pursuit.

"Starn chase," said the captain, "but we'll hev him in half an hour, yaou mark. Let's see: I was abaout tellin' of my first so'd fish. Well, as I was a-sayin', I was powerful green—spent most o' the time a-mumin' araound the farm. After a time I got a-coastin' on father's vessel, and one summer he couldn't git no work for her, so I says, 'Father, let me take the schooner and go a-fishin'.' She bein' insured and hogged (broken-backed), he agreed. So the next night we was off and baound up to George's Banks, not a-livin' soul of us ever havin' been there afore, and all young and fresh as a spare room. We had poor luck on the Banks, for the reason that I found aout after, —we hed never struck the Banks at all, and hed been a-cod-fishin' off bottom. Howsomever, we worked up the coast,— hevin' no charts, from pint to pint, and from light to light, fur several days; then, water runnin' low, we hailed a lumberman baound daown, who gave us the bearin's of a place tew north'ard and east'ard. We kept up till we struck a bay and run abaout twenty miles; then, night a-comin' on, we come tew anchor abaout three miles from shore, in abaout five fathom, and, everythin' bein' snug, all hands, four of us, turned in in ham-mocks slung in the hold, that bein' the coolest place. Well, the next mornin, John Hanson sings aout, 'D'ye hear that, Sam?' 'What?' says I. 'Hogs,' says he. 'Go 'long,' says I. But, sure enough, there was the greatest gruntin' and squealin' yaou ever heard.

"'Queer country this,' says John; 'hogs swim aout tew ves-sels.' I thought so myself, but I didn't let on. The gruntin' was gittin' wuss and wuss, so I reached aout my head, and it

PLATE VIII.

SOUTH AMERICAN CAT FISH CRAWLING ON DRY LAND.

was jest light enough tew see the snout of a big hog lookin' daown the hold. That settled it. When it came tew a drove of hogs swimmin' off tew a vessel and climbin' aboard, it was sartin time tew turn aout. I rolled aout first, and daown I went; the old schooner had a lurch tew port on her, so you couldn't stand up. 'We're going daown,' says Hanson, rollin' aout. 'Nonsense!' says I; 'she's on her beam ends,' and with that we all made a break for the hatch and come on deck. I ain't no hand tew be took aback, but for a spell we sot on the combin' of the hatch in a regular sog. She was high and dry, with fifteen fathoms of cable aout, and hogs a-feedin' and rootin' araound the starn post, and some on 'em on deck. Ye couldn't see water fur ten miles away. We was powerful sot back. 'Is she insured?' says John. 'I believe she is,' says I. 'Well,' says John, somethin's happened,—'arthquake, or a powerful parch,—and I'm fur strikin' in.' So we packed up and walked ashore through the mud, nigh on tew three miles. At last we come across a man diggin' clams! 'What station is this?' says John Hanson. 'That's Monckton over there,' says he. 'What's the flat called?' says John. 'Oh,' says the man, 'this ere's the Bay o' Fundy, and if yaou're a-goin' to Monckton ye'd better shake a leg afore flood-tide.' 'High tides, here?' says John, kind o' knowin'-like. 'Tolerable,' says the man; 'a like o' fifty feet or so.'

"Well, three days later, we was tew home and aout o' commission. We'd never heard of no big tides, and I reckon this is the first time the tellin' o' that ere master parch has been let aout. Dretful green!—dretful!" And Captain Sam leaned over the binnacle, grew a deeper red, and laughed spasmodically, gradually regaining his former equilibrium.

"And the sword-fish?" I asked.

5

"Oh, yes, the so'd-fish. Well, we found it alongside, left by the tide; that's what the hogs was after."

During this recital we had been gaining on the keg, and now, at a word from the skipper, the dory that was bounding along astern was hauled alongside, the painter tossed in, and the entire port watch, numbering two souls, and the passenger, as volunteer, literally tumbled in, and were soon moving toward the jumping buoy, that was now almost stationary.

"Yaou clap on to the keg, mister," said John, the veritable Hanson of the great parch, "and we'll lay you alongside."

A few sturdy strokes, and the keg swung by, and was secured; oars were jerked in, and not a moment too soon was the line slipped into a crotch in the bow, for the gamey fish, feeling the line tauten, leaped into activity, jerking the volunteer ignominiously among the bailers.

The cry "starn all," in a literal sense, was not needed; we were all there, and with bow half under, were headed out, taking everything as it came. A fair sea was running, and soon our small craft started an opposition wave, that, curling several feet ahead, seemed leading us to victory or a capsize.

"She won't heft many o' them," said John, as a big wave came slashing in upon us. "We'll have to get in that slack." And seizing the line, he passed it aft, and the struggle, three to one, commenced. To gain a fathom of line was hard work, the fish now using desperate efforts, making long surges to the right and left, or cleaving the waves its entire length, in vain endeavors for freedom. In twenty minutes we were in sight of the monster, and, with a shout, all hands laid on, and the game was alongside.

"Pass the line astern!" shouted John, and in the struggle down we went with a crash, quickly climbing to windward to

avert the catastrophe threatened by the line fouling in a row-
lock. For a moment we were upon a dizzy height—on the
upper rail ; then the line slackened and cleared, and with a rush
was passed to the scull-hole in the stern.

"Lay on naow, hard !" shouted some one, and lay on we did,
knee-deep in the water shipped during the flurry. One good
pull all together, and the line was "chock up," and the fish fast
astern. The green hand venturing to prospect the field, the
ugly sword came flying over the dory, creating a "down-
bridge" movement. The sinewy form bent in great curves,
straightening out with extreme rapidity, making slashing blows
against the boat, while the sharp tail quivered and glanced, cutting the water like a knife.

To the disinterested observer it would have been an exhilar-
ating spectacle. The volunteer, being in the front row, was
moved from not altogether humanitarian feelings, to cut the
rope ; but this now proved unnecessary : the game was up.
The gallant swordsman was weakening ; the slashes, rushes
and bounds became less frequent, and finally, as the sharp sword
came against the gunwale, John Hanson caught it with a round
turn and securely lashed it high out of water : the gamey fish
was conquered.

"It's kind of a question in my mind," screamed Captain
Sam, as the Thumbscrew rounded to, " whether yaou caught the
fish, or he caught yaou."

The painter was thrown aboard, and a running bow-line
passed around the fish's tail, and as the dory dropped astern the
Xiphias gladius was hoisted aboard, the schooner filling away to
bide the call of the fresh man who now went aloft. The fish
proved to be nearly nine feet long, with a perfect sword, so
well tempered that it had received no damage from the heavy

blows against the dory. Every feature betokened speed and activity ; its whole appearance was rakish, that of the privateer ; the dorsal was tall and graceful, the tail keeled, the lower jaw sharp, and the back a rich bluish black grading off to a clear silvery white below.

" He's good fur eight dollars," remarked Captain Sam, as we went aft after relieving the monster of several parasites—*Penella filosa*, etc.—that infested him.

" Yaou see," he continued, " they're a kind of mackerel—belong to the same family—and there's always a big demand fur 'em. Strike a ship? Well, I should say so. I shipped several years ago on the Maria Jane from Gloucester, and while a-mackerel-fishin', we was struck by a so'd-fish. The first thing we knowed the wheel wouldn't work, and, on lookin' over, there was a so'd a-stickin' in between the rudder and the post, broken off short, and it took us a couple of hours to git it aout. I reckon it kills 'em in the long run. It's a common thing fur us to strike 'em without any so'ds. Sometimes they break 'em in whales, or agin' vessels ; and I've hauled 'em in when their heads was all mud, showin' how they'd rushed agin' the bottom, and perhaps broken it off in that way. But when the so'd's gone they're always poor ; so it makes me think they don't feed without it. The habit's so strong in 'em to strike a fish, that they think they must do it. Why, I've tossed a dead porgy to one and seen him knock it up and daown, jist like a game of bat and ball, afore he'd touch it, and then kind of slide under and come up and take it. I've caught 'em on a hook and line, and in a herrin'-net,—and bought a new net, tew, with the very money I got fur 'em ; obliged ter do it. But there's one curious thing ; yaou can't find a so'd-fisherman on the coast of Maine, thet, winter or summer, ever see a young one. There's a good many folks disbelieve they hev young; but I've heard tell haow they breed on the other side of the water."

The captain's idea was the right one, and, on careful inquiry, I could not find a fisherman on the shore that had ever seen a young sword-fish. They belong to one of a number, including the horse-mackerel, that perhaps cross the Atlantic, depositing their eggs, for some unknown reason, on European shores, or in unknown waters. In the Mediterranean Sea, the females approach the shore in the latter part of spring or the first of summer. The white compact flesh of the young fish is held in great esteem, that of the adult resembling the tunny, and preferred to the sturgeon or halibut, which it somewhat resembles in flavor.

The young sword-fish differ greatly in appearance from the adult. The young of the genus *Histiophorus*, when about nine millimetres long, have, according to Gunther, jaws of equal length, armed with sharp teeth; the dorsal and anal fins are a low-fringe; and the ventrals make their appearance as a pair of short buds. When fourteen millimetres long, the young fish has still the same spines on the head, but the dorsal fin has become much higher, and the ventral filaments have grown to a great length. At a third stage, when the fish has attained to a length of sixty millimetres, the upper jaw is considerably prolonged beyond the lower, losing its teeth; the spines of the head are shortened, and the fins assume nearly the shape which they retain in mature individuals. Young sword-fishes without ventral fins (*Xiphias*) undergo similar changes; and, besides, their skin is covered with small rough excrescences longitudinally arranged, which continue to be visible after the young fish has in other respects assumed the mature form.

The sword-fish industry on the Maine coast and south of Cape Cod is an important one, employing a large number of men and boats. The pursuit is one of the most ancient on record. Strabo mentions it as having been followed in the days of Ulysses. Pliny associates the fish with several others as "suitable to use as salted fish," and Rondelet describes it under the name "*Poisson Nommé Empereur.*" In the Greek it was *Xiphias;* in the Latin, *Gladius;* the Italians call it *Pesce spada;* while in France it is known as *Héron du mer,* and *Poisson Empereur.*

In the Straits of Messina the fishery is of great importance, numbers of men from Messina and Reggio being employed. The boats at night are lighted with huge flambeaux, which are supposed to attract the fish, while a man stationed aloft upon a single mast announces the approach of the game. The Sicilian fishermen, as well as those of Reggio, chant a peculiar jargon during the chase, supposed to be a sentence in Greek, to charm the fish within reach of their harpoons, the common belief being that if the fish hears a word of Italian he will dart to the bottom. Kircher took down the words, and found that they were more like Hebrew; and he suggests that they are a remnant of the ancient Phoenician tongue.

SAILOR-FISH.

There are three genera well known : the common *Xiphias,* a species of which is found in South-American waters twenty-five feet long, the bill—or bayonet—fish (*Tetrapturus*), and the great sail—or sailor—fish (*Histiophorus*) of the Mediterranean and Indian Ocean. The last attains, near Ceylon, a length of twenty-five or thirty feet, the enormous dorsal fin, which is in this locality retained in the adult fish, often being ten feet in height, and presenting a strange appearance, rushing along out of water, and scintillating with blue and golden tints. (*Plate XI.*)

Many interesting cases illustrating the pugnacity of these swordsmen have been recorded. Sir Joseph Banks cites an instance where the entire sword was driven into the solid wood of a ship ; and in another case, in which the sword had penetrated the copper sheathing and three and a half inches of solid oak, competent judges estimated that to drive a pointed bolt of iron of the same size and form to the same depth would require nine or ten blows of a hammer weighing thirty pounds.

Professor G. Brown Goode, in his material for the history of the sword-fishes, enumerates various instances showing that, far from being unusual, these attacks of sword-fish are of yearly occurrence.

In the London *Daily News* of December 11th, 1868, the following statement is made :

"Last Wednesday the Court of Common Pleas—rather a strange place, by-the-by, for inquiring into the natural history of fishes—was engaged for several hours in trying to determine under what circumstances a sword-fish might be able to escape scot-free after thrusting his snout into the side of a ship. The gallant ship Dreadnought, thoroughly repaired, and class A 1, at Lloyd's, had been insured for three thousand pounds against all the risks of the seas. She sailed on March 10th, 1864, from Colombo, for London. Three days later, the crew, while fishing, hooked a sword-fish. *Xiphias*, however, broke the line, and a few moments afterward leaped half out of water, with the object it should seem of taking a look at his persecutor, the Dreadnought. Probably he satisfied himself that the enemy was some abnormally large cetacean, which it was his natural duty to attack forthwith. Be this as it may, the attack was made, and at four o'clock the next morning, the captain was awakened with the unwelcome intelligence that the ship had sprung aleak. She was taken back to Colombo, and thence to Cochin, where she was 'hove down.' Near the keel was found a round hole, an inch in diameter running completely through the copper sheathing and planking.

" As attacks by sword-fish are included among sea-risks, the insurance company was willing to pay the damages claimed by the owners of the ship if only it could be proved that the hole had really been made by a sword-fish. No instance had ever been recorded in which a sword-fish had been able to withdraw his sword after attacking a ship. A defence was founded on the possibility that the hole had been made in some other way. Professor Owen and Mr. Frank Buckland gave their evidence, but neither of them could state positively whether a sword-fish which had passed its beak through three inches of stout planking could withdraw without the loss of its sword. Mr. Buckland said that fish have no power of 'backing,' and expressed his belief that he could hold a sword-fish by the beak; but then he admitted that the fish had considerable lateral power, and might so 'wriggle its sword out of a hole.' And so the insurance company will have to pay nearly six hundred pounds because an ill-tempered fish objected to be hooked and took its revenge by running full tilt against copper sheathing and oak planking."

In 1875, the Gloucester steamer Wyoming was similarly struck, so that the men had to take to the pumps to keep her free; and a like experience was that of the English brigantine Fortunate. Her captain reported that while on his passage from Rio Grande, his ship was struck by a large fish, which made the vessel shake very much. Thinking it had been merely struck by the tail of some sea-monster, he took no further notice of the matter; but, after discharging cargo at

PLATE IX.

DRY BURROW OF THE LUNG FISH.

Runcorn and coming into the Canada half-tide dock, he found one of the plank ends in the stern split, and on closer examination discovered that a sword-fish had driven his sword completely through the plank, four inches in thickness, sending the point of the sword nearly eight inches farther. The fish, in its struggle, broke the sword off level with the outside of the vessel, losing nearly a foot of the weapon. There can be no doubt that this somewhat singular occurrence took place when the vessel was struck as Captain Harwood describes.

The experience of Captain Wm. Taylor, of Mystic, with one of these monsters, is to be remembered. He started from that place in October, 1832, on a fishing voyage to Key West, in company with the smack Morning Star. They were off Cape Hatteras, the wind blowing heavily from the northwest, and the smack under double-reefed sail, when at ten o'clock in the evening they were struck by a "woho," which shocked the vessel all over. The smack was leaking badly, and they made a signal to the Morning Star to keep close by them. The next morning they found the leak, and both smacks kept off to Charleston. On arrival they took out the ballast, hove her out, and found that the sword had gone through the planking, timber and ceiling. The plank was two inches thick, the timber five inches, and the ceiling one and a half inches white oak. The sword projected two inches through the ceiling, on the inside of the "after-run." It struck close by a butt on the outside, which caused the leak. They took out and replaced a piece of the plank, and proceeded on their voyage.

On the return of the whale-ship Fortune to Plymouth, Mass., in 1827, the stump of a sword-blade of this fish was noticed projecting like a cog outside, which, on being traced, had been driven through the copper sheathing, an inch-board undersheathing, a three-inch plank of hard wood, the solid white oak timber twelve inches thick, then through another two and a half-inch hard-oak ceiling, and lastly penetrated the head of an oil-cask, where it stuck, not a drop of the oil having escaped.

In a calm day in the summer of 1832, on the coast of Massachusetts, a pilot was rowing his little skiff leisurely along, when he was suddenly aroused from his seat by a thrust from below by a sword-fish, who drove his sharp instrument more than three feet up through the bottom. (*Plate XII.*) With rare presence of mind, with the butt of an oar he broke it off level with the floor before the fish had time to withdraw it. Fortunately, the thrust was not directly upward. Had it been so, the frail boat would have been destroyed.

The secret of these attacks upon vessels may perhaps be explained by the fact that the ship is mistaken for some enemy.

On a long reach to the eastward, yet within hearing of the groaning buoy off Boon Island, we captured another sword-fish, weighing two hundred and fifty pounds, and in the course of the day four fine specimens were added to the number, deciding our skipper to find a market; and, after a consultation with John Hanson, who was a third owner in the Thumbscrew, in which the fall of a tossed penny seemed to shape the course, the sheets were slackened off, and wing and wing the sword-fisherman bore away, by the Isles of Shoals, for Gloucester town.

CHAPTER VII.

FINNY LIGHT-BEARERS.

AMONG the most interesting light-givers of the abyssal depths of the ocean are the fishes. Some have remarkably large eyes, perhaps emitting light themselves; others have curious phosphorescent organs about their heads and bodies. The late Professor Willemoes-Suhm saw directly the phosphorescence of the curious fish *Sternoptyx*, and records it as a most wonderful exhibition; while of another he wrote, " It hung in the net like a golden star." Another striking light-giver is the brilliant lampfish *Scopelus resplendens* (*Plate XIII.*); on its sides and various parts of the body are numbers of curious round pearly spots that emit a phosphorescent light, while upon the forehead a brilliant blaze is seen, scintillating as it rushes along, beaming like a miniature headlight of a locomotive, and where a shoal of them is seen, we can imagine them the *ignus fatui* of the sea; the lights moving up and down, now here, now there, appearing and reappearing in a strange ghostly manner. When the water about them is phosphorescent of itself, the scene is still more striking: flashes of flame dart across the field of vision, and here the outline of the fish is seen in fire that is lost in a brilliant glare as the finny light-bearer darts away, followed by a train of radiance.

In another species, the saury, *Scopelus Humboldti*, we find the same singular phosphorescent spots that have occasioned so much discussion among naturalists. Although it is only quite lately

75

that these have attracted any special attention in other fishes, their presence in *Scopelus* was long ago noted by various observers.

Very similar in general appearance to the *Sternoptyx* is the *Argyropelecus* (*Plate XIV.*), a quaintly formed fish from the waters of the Mediterranean. If we examine it, as did Drs. Ussow and Leydig, we shall find that there is one spot in front of the eye, and behind it two; six smaller ones are found on the gill-membrane, and six larger ones on the throat; by the gill-cleft we see four, the largest on the body, of which two lie in front, and two behind the cleft. At the side of the body, and close to the abdominal profile, there are twelve, which decrease in size from before backward; and above these there is a second row of six, all of which are of very much the same size. Between the ventral and anal fins there are again four, and behind the anal fin there are six others, of which the smallest are in the middle. Just in front of the tail fin we find the last quartet of these organs. This gives us fifty-three in all, and as they are paired, we find one hundred and six of these comparatively large spots on the body of this small fish.

Professor Leydig entertains the belief that the spots are not alone luminous, but may be divided into three groups: (1) eye-like organs, (2) mother-of-pearl-like organs, and (3) luminous organs. In the fish *Chauliodus* there are thousands of these spots; enough, if they all give light, to mark the creature against the water in lines of fire, while in the *Stomias*, with its snake-like body and frightful head, we can imagine rows of lamps, serving as warning-lights, if indeed its hideous appearance was not enough to repel all curious fishes.

An interesting example of the large-eyed light-givers, is the Ipnops, a new fish found by the "Challenger," its great orbs being described as blazing with phosphorescence in the night. One of the most familiar light-givers, however, is the moon-fish, or sun-fish, of our coast. In the Mediterranean they are very common, and at night present a remarkable appearance, moving along like gigantic globes of light, or resembling the reflection of the moon upon the water. When numbers of them swim along together, the huge dorsal cutting the water, that breaks itself into ripples of molten silver, the scene is described as being magnificent in the extreme.

The curious faculty of phosphorescent light appearing both in life and death in the same subject, is nowhere so well shown as in the fishes, where many that have not been light-givers during life now seem invested with this light-emitting power. In the mackerels, pompinos, jacks, etc., this is especially noticeable. If they are touched, the luminous substance, which seems to be oily, comes off upon the hands, so that they appear as if rubbed with phosphorus. A German scientist says if the fish are placed in water the light is communicated to it, showing that it is some peculiar secretion—a seeming combustion without heat, at least, the most delicate instruments known to science fail to detect it.

The phosphorescence of dead fish is not confined to salt water, as the carp has been seen to gleam with light. During the summers of 1880 and 1881, the daily papers from time to time treated their readers to selections from the Flying Dutchman, revised, however, to suit the times. The inhabitants of the eastern end of Long Island and "down Block Island way," according to these romancers of the pen, were in mortal terror of a phantom ship that was often seen cruising about the vicinity. The accounts were denied by many readers, and the war of words succeeded in drawing out a reputable citizen of

Rhode Island, who affirmed that he had observed the phantom. He described her as a full-rigged ship, and had seen her every year for a long time; sometimes she hove in sight off the harbor at dusk, rushing on toward shore, and when about to dash against the rocks, would disappear as suddenly as she came. This much respected but credulous gentleman was a firm believer in the phantom ship, for the very good reason that he had seen her beating to windward when there was not a breath of wind. "The first time I saw it," he said to an acquaintance, "our yacht was lying off Gardiner's Bay one warm summer night, when one of the men, an old sailor, standing by me, said, 'There's a sail beyond; the wind's a-freshening outside.' It was nearly dark, but out on the sound appeared a vessel that every moment came nearer. There was not a breath of wind, yet on she came, the foam piling up under her chains, and everything standing. 'That aint a vessel,' said the old man, after we had gazed at the strange sight awhile in wonderment. The great white mass had come on so rapidly that she now seemed within a fourth of a mile of us. 'What else is it?' I called, as he moved away. 'It's a barque, and she'll run us down!' he answered. 'Ahoy, there!' I hailed to the man on the for'castle, rushing forward at the same time to warn them; and springing into the rigging, I swung out and hailed the craft, 'Ship ahoy! Bear away! Ship ahoy!' But it was too late. I clung riveted to the spot, and with another cry to the men to jump for her chains, I stood ready to follow suit, thinking it our only chance. On she came; I could almost feel the spray from her cut-water, when I felt a tap on my shoulder, and the old tar spoke up, 'I told you that 'ere wasn't a living ship.' And, sure enough, she was gone; whether up or down, in the air, or where, I can't say; but I saw her as distinctly as

I ever saw anything in my life, and have seen the same thing again, many and many a time."

One peculiarity about the phantom ship is that she always appears at or near night, and there is no reason to doubt that something in appearance resembling a ship has been seen by many along the New England coast, and the phantom vessel, in the author's opinion, will continue to cruise about and be seen along shore by those who stay out o'nights, until the last run of Menhaden. Every season from eight to ten millions of these fishes are caught in Long Island Sound, their oil amounting to six gallons per thousand. That these fish emit an oily substance every fisherman knows; while many others who go down to the sea in ships not as professionals, must have seen the immense "slicks" of oil, often miles in extent, left upon the water by them. Even if each fish emits an infinitesimal amount of such matter, the effect of millions of such emissions rising to the surface by natural means or from wounded or dying fish, as the school moves along, might, under certain conditions, become luminous, and, to the imagination of lookers on, assume the guise of the famous Dutchman.

The phosphorescence of crabs was first discovered by Sir Joseph Banks on his voyage to Rio Janeiro from Madeira. A small crustacean (*Cancer fulgens*) was taken aboard one evening, which gave a wondrous exhibition of its light-emitting power. Its entire surface seemed bathed with a white flame, that flashed and sparkled like living fire, and so resembled it that later one of the sailors picked the crab up, thinking it a coal that had rolled out of the galley fire.

Mm. Edoux and Toulezet, two French naturalists who made a scientific voyage around the world, observed that certain small phosphorescent crustacea sometimes secrete a peculiar phosphorescent matter, and that when they are irritated they send forth magnificent flashes of light. These gentlemen collected a certain quantity of the phosphoric substance, and found it to be yellowish, viscous, and soluble in water, communicating its luminous property to this liquid, but only for an instant or two. It lost its luminosity when it had been separated for a few moments from the body of the animal.

The deep sea crabs have phosphorescent eyes, especially *Geryon tridens*, *Gonoplax*, *Dorynchus*, and *Munida;* the latter being particularly noticeable for the balls of fire into which the tips of their stalked eyes seem to have been converted. In some the eyes have totally lost their proper functions, and assumed those of phosphorescent organs.

The little Cyclops of our fresh water ponds, that form so beautiful an object under the microscope, have been observed to gleam with a silvery light. A remarkable spectacle in which these single-eyed crustaceans took part was observed near Bloomington, Ill., a few years ago in a rain-storm during the night. Before daybreak the gutters, roofs and streets were found to be covered with quantities of these little creatures that appeared like molten metal, gleaming and glowing with a wonderful light. They had been caught up by a whirlwind from some distant point, and deposited here by the rain. In the daytime they presented the appearance of minute yellow and reddish specks, some of them possessing an extremely brilliant and vivid coloring.

Along our sea-shores we may often see under the rocks, clinging to the eel-grass, in some pool left by the tide, gleaming spots that move about in erratic courses; now many collecting together, then breaking up into small patches of light which in turn separate again. They are curious crustaceans, known

PLATE X.

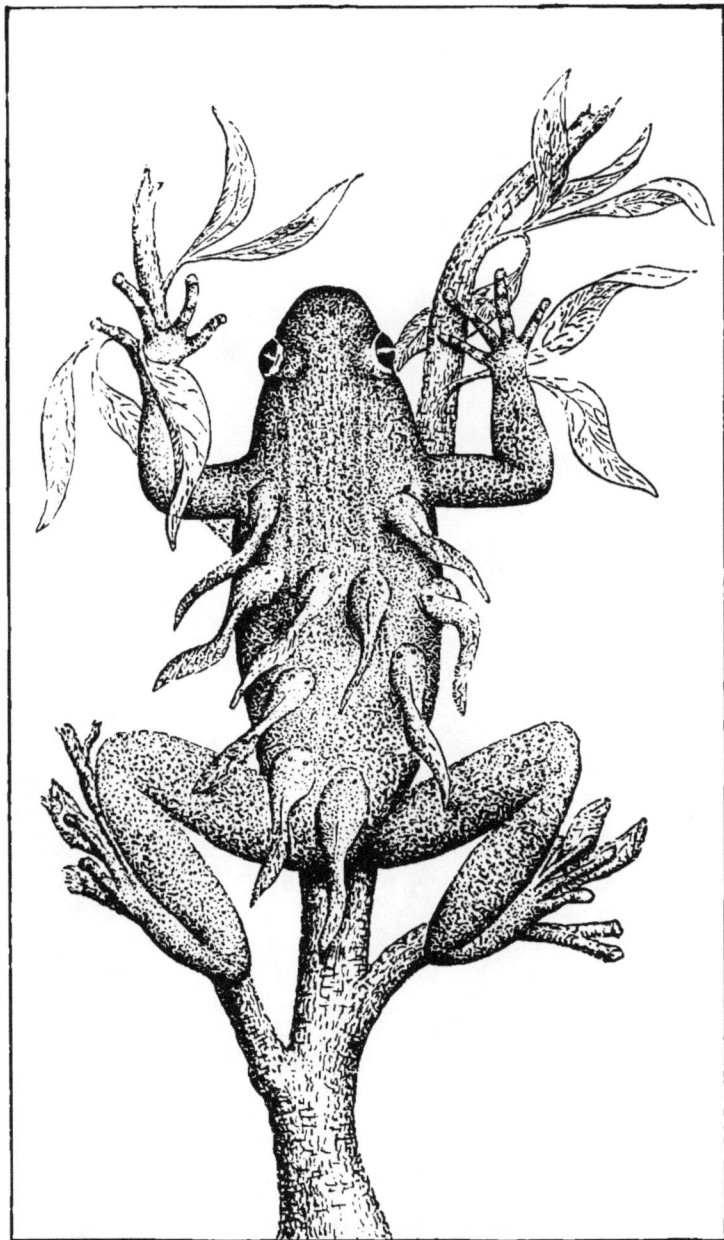

MARTINIQUE TREE TOAD WITH YOUNG CLINGING TO ITS BACK.

scientifically as the *Idotea phosphorea*. We shall find that they are usually spotted or entirely a bright yellow, a peculiarity that the reader will notice in nearly all light-giving animals.

In the Arctic regions beautiful lights have been observed that are due to a minute crustacean. Lieut. Bellot first witnessed it in the North American polar regions, and Nordenskiöld refers to it in his voyage of the Vega, the most brilliant displays being seen at Mussel Bay. He says: "If during winter one walks along the beach on the snow, which at ebb is dry, but at flood-tide is more or less drenched through with sea water, there rises at every step an exceedingly intense, beautiful bluish-white flash of light, which in the spectroscope gives a one-colored labrador—blue spectrum. This beautiful flash of light arises from the snow that shows no luminosity before it is stepped upon. The flash lasts only a few moments, but is so intense that it appears as if a sea of fire would open at every step a man takes. It produces indeed a peculiar impression on dark and stormy winter days (the temperature of the air was sometimes in the neighborhood of the freezing point of mercury) to walk along in this mixture of snow and flame, which at every step one takes splashes about in all directions, shining with a light so intense that one is ready to fear that his shoes or clothes will take fire."

The cause of this phenomenon is a little crustacean, *Metridia armata*, that much resembles the cyclops, and the great changes of temperature to which it is subjected in the snow sludge seem to have no effect upon it.

Many of the worms common on our own shores are interesting light-givers, some rising to the surface at night, presenting a brilliant appearance as they wind their way over the sea. The most striking forms are found in the families *Polynoidæ*,

6

Syllidæ, Chaetopteridæ, and *Polycirrus.* The first emits a greenish light at the attachment of each scale. In the second the surfaces of the feet seem to be the luminous points. In the third the light appears about the tenth joint, while the *Polycirrus* is a veritable fiery worm, its entire surface gleaming with a vivid bluish light.

To observe these and other small luminous animals, as the *Noctilucæ,* etc., a night-laboratory should be set up upon some rocky point where pools are left at low tide. A dark night should be selected, and with a good microscope and a bull's-eye lantern, a very enjoyable evening may be passed. The weed clinging to the rocks and partly submerged, will afford abundant material, as when a wave recedes, the pendent bunch will often be seen to fairly blaze with the luminous forms. There is an additional zest added to the investigations in this direction from the fact that the subject of phosphorescence is still in its infancy as regards explanation, and comparatively nothing is known concerning the why and wherefore of the phenomena, that appear in life and death, in growth, and in decay.

CHAPTER VIII.

OLD FRIENDS.

OUR pets were perhaps more remarkable for their variety than for their display of intelligence, which was not strange, considering that, as a rule, they were not selected from the high and exalted ranks of society, that includes the monkeys and parrots, but formed a part of the great class popularly called the lower animals. Perhaps some of them would hardly be thought to come under the head of pets when I confess that in several instances they were not in confinement at all, but had the range of the great coral shoal, only restricted by the line of breakers that pounded upon the outer fringing reef around our tropical home. In the shallow lagoon were numbers of what we called heads of coral, enormous oval masses of astrea and meandrina, sometimes four feet across and three or four feet in height. Many of these were hollowed out from various causes; the coral polyps perhaps being killed upon the upper surface at extreme low tides, various worms helping on the process of disintegration, until finally the head was a gigantic coral vase peopled by numerous strange forms.

Now as many fishes, crabs, and other animals exhibit a partiality for their homes just as we do, I could almost always find the same animals on succeeding visits; and as these heads were visited many hundreds of times during the years spent on the reef, you can well imagine that we became very friendly.

But who, you will ask, were our friends? First there were

the craw-fishes. They occupied the first floor, and lived beneath the edge of the heads where they had excavated holes in the sand for the purpose, and all around the borders you could see their whip-like feelers moving nervously to and fro. They were good and valued friends of the coral, as by making their homes there they prevented the fatal inroads of sand that would have destroyed the delicate polyps.

In the vase, or natural aquarium, would be found numbers of richly tinted anemones, several young craw-fish, sea-eggs (black echini, with long needle-like spines), a variety of star-fishes, curious blue crabs, and above all, numbers of small fishes, from their beautiful colors, known as angel fishes. The latter in their movements were extremely graceful, and they would often allow us to approach within a few feet of them. In deeper water, in swimming along the face of a branch coral bed, I have been so near some of these wondrous birds of the sea that I could almost touch them, and of course could see them distinctly, although we were both perhaps twenty feet below the surface.

One day in visiting a favorite coral head, I noticed several small cow-fishes come out of a hole, and as they were within easy reach, I extended my hand slowly, when to my astonishment they met me half-way, examining my fingers curiously, being especially attracted by the nails at which they delicately nibbled. I then opened my fingers, and they swam between them without showing the slightest fear, and as I was very careful not to alarm them, we were soon on terms of close friendship. I next ventured to lift them in my hand, and as I raised one up nearer the surface, its brother or cousin would follow along as if determined not to be left behind. Finally, on another occasion, I lifted one from the water, where it presented a comical appearance, not being able to frisk about like other fishes, as its body was encased in a hard armor, only the tail and fins being movable. The position was not a natural one, so I soon returned the little creature; but even this rough treatment did not cause them to lose faith in me, and for a long time I visited the head, occasionally treating them to soft bits of craw-fish (*Palinurus*) and other dainties that they were fond of.

One of the commonest forms upon the reef was the hermit-crab. Several kinds were known: one that lived entirely in the water, a great red-clawed fellow that hauled about a shell of the horse conch, *Strombus gigas*, and others much smaller that were always found on shore. Almost every empty shell upon the sands was occupied by one of these curious little fellows, as they are so formed that an artificial home is necessary to their safety, and this is changed several times a year, or as fast as they outgrow it.

The land-crabs were often very ornamental, having red and purple claws. I generally kept several in my room, and to render them still more attractive ground off the rough outside of their shell, which was usually a Top-shell (*Trochus*), exposing the rich pearly interior, so that in houses of pearl they presented an appearance altogether gorgeous. The clank, clank, of their shells could be heard at any time about the room, and I found them in all sorts of strange places. One day in lifting over a board pile to find some crabs, this being a favorite place for them, as well as for scorpions, I came upon a large old-fashioned clay tobacco pipe, stemless and evidently cast aside. Wondering how it came in such a position, I picked it up and solved the mystery, for snugly coiled up within was the quaint old fellow shown in the accompanying picture, suspended from a branch. (*Plate XV.*) The old pipe-bowl had been selected as a home, and a very good one it proved.   This unique example prompted us to adopt the crab as a special pet, and the position shown in the picture was a result of its venturesome nature, as it took the liberty to pass out of its bounds and climb a neighboring shrub.

Diogenes, as we named it, became very tame.   Its actions were exceedingly grotesque; especially so was its passage up the corner of a bookcase, on which it crawled daily to get a drink of water which was placed there upon a ledge.   This species usually occupied a trochus shell about half the size of one's closed hand; but our pet had a freak for something out of the common, and adopted the old pipe.   We took several of these hermits North, where they lived over one season, moulting twice, and became tame enough to eat from one's hand.

Later we found another hermit in a pipe-bowl, this time a marine form, much smaller and less interesting, as it was extremely timid, withdrawing into the pipe at the slightest warning.

Several gophers, land turtles, were found upon the key and kept as pets, but they remained unimpressed by all acts of friendship, as did a young hawk-bill turtle that we rescued from the tentacles of a physalia.   The hawk-bill would feed from my hand, but always resisted any further advances.

The most interesting of our reptilian pets were the recently hatched green and loggerhead turtles.   These curious little creatures were often caught coming from the nest, and at one time I had several hundred, each about an inch and a half long, loggerheads in miniature, in our study at the same time.   This apartment was built at the edge of the water, a door leading from it into an inclosed aquarium, and it was interesting to note how well their instinct served them, as when both doors were closed, and the little creatures were turned about repeatedly to confuse them, they would turn as soon as placed upon the floor, and head directly for the door that led to their native element, clustering about it in groups, and when it was opened, tumble headlong into the welcome water.

Besides these humbler pets, there were a number of goats, one of which deserves mention for her remarkable intelligence and distinctive traits. As a pet she was like a dog, following the members of the household about, and insisting upon remaining in the house, showing a decided preference for the library, where she contented herself with the bindings of books. In fact, Bon was an expensive luxury. Numerous valuable articles, such as lace collars, important papers, Panama hats, etc., being considered by her as choice delicacies. One of the last of her depredations that I remember witnessing, was seeing her leave my side and rush at a box that stood among the trees, and seize a long, eight-page letter that some one had just written and left for a moment.

Bon was thoroughly aristocratic; she would not associate with the other goats; and to show how she recognized the power of rank, I will cite an instance. In her prime the fort was garrisoned by a regiment of volunteers. Bon was a great pet among the officers, but had an intense dislike for a private soldier, for the simple reason that whenever she attempted to pass the sally-port, they drove her back and occasionally caught her.

One day, however, an idea entered Bon's head that completely outwitted the men. When she wished to pass the sentry she would wander in that direction, nibbling along by the palm trees, or perhaps lie down until an officer came in sight, when she would quickly join him and follow along, and while the men and sentinels were standing at attention and saluting their superior, she would pass through the gate in safety, close at the heels of her protector. This ruse was adopted nearly every day, and became thoroughly appreciated by the men. Finally it was taken up by other goats of the garrison, who took advantage of the situation and followed Bon, and the men standing at attention would have difficulty in repressing a smile, as the officer of the day passed by, followed by Bon with her glistening brass-tipped horns and shining black and white coat, who formed a gallant leader to the herd of goats behind.

Bon was not devoid of humor, and was possessed of the most exuberant spirits, one of her tricks being to rush at the side of a house, plant her four hoofs firmly on it, several feet from the ground, and leap away with extraordinary agility. This performance was repeated time after time, by bidding her do it, and when applauded she would toss her head as if decidedly pleased.

When we finally left the fortress, Bon was domiciled at an island near Havana, where we afterward learned she lived to a good old age. She had been a faithful friend, and supplied us with all the milk we had, as there were no cows. But it must be admitted it was an ungracious gift, as it generally took three persons to milk her; one to hold her head, another her hind feet, while the third milked her. Even then she managed to upset all three and the pail, at times.

Among the pets that from time to time have taken our attention, might be mentioned an iguana, an American chameleon, and divers horned toads that formed interesting specimens in which to study the changes of color described in the chapter on mimicry. At one time we had forty different spiders domesticated in glass bottles that were filled with their webs.

Snakes and toads were not the least among the dependents of our Northern home, and one of the latter was especially esteemed

PLATE XI

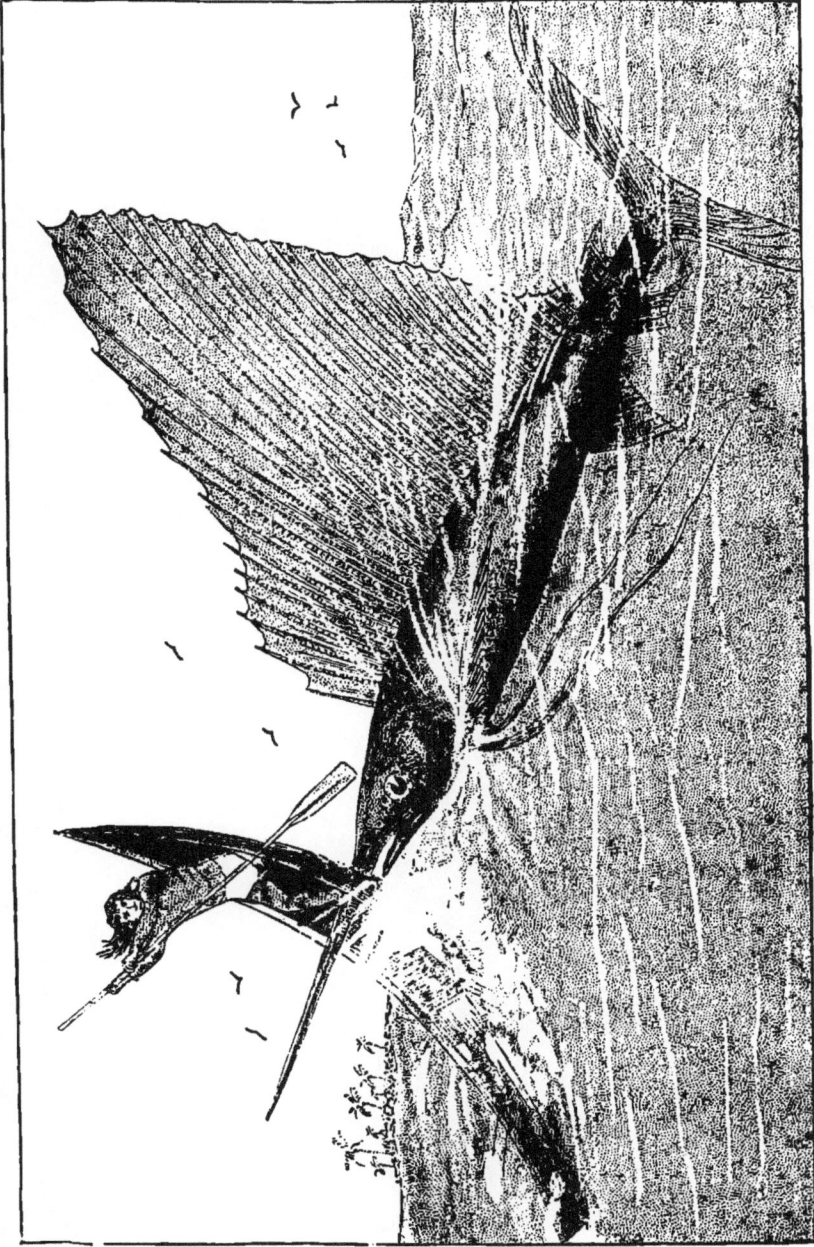

THE SAILOR FISH OF CEYLONESE WATERS.

for its skill as an architect. It had selected for its summer house a location over which a stream of water ran every time a shower came up, and, consequently, the pit became flooded. Several lessons of this kind evidently made his toadship put on his thinking cap, for one day I discovered that the hole had been enlarged by an ingeniously devised chamber, or tunnel, that led up and away over the pit, so that the latter was now a safety drain for the real abode, in which the toad resided in perfect security.

Our flying-squirrel, notwithstanding his fancy for forming a nest out of a certain silk dress, was a most entertaining little fellow. Often has he sat on my inkstand, pretending to handle a nut, as I wrote, and endeavoring to distract my ideas by the twinkling of his bright black eyes; now suddenly darting up my sleeve and into a pocket; now on my head, or taking a flying leap to the floor where he scampered around the room in a wild and gleeful race.

One day he crawled into the lining of a lady visitor's bonnet, and went to sleep, only awakening when the owner had nearly placed it on her head, creating, as you may think, a deal of excitement. Our friends were not alone the victims of this mischievous creature. One night after retiring I was aroused by a loud roaring in my ears, and as I had been taking quinine, I presumed that was the occasion of the sound, and that I had taken an overdose. But as it momentarily grew louder, and I imagined that my head was whirling about, somewhat alarmed I got up, when it instantly ceased. Astonished I struck a light, and going to my pillow, there in the pillow case, just where my ear had been, I found Peter, for so we called him, still making the curious whirring sound flying-squirrels often make when about to go to sleep.

Among our fishes, tri-tailed Japanese gold fishes, dace, pickerel and cat-fishes, the sun or pond-fish proved the most interesting. One especially I kept for nearly three years. At first he was most pugnacious, attacking and destroying other fishes and animals with the greatest fury, until finally he had the tank entirely to himself. He soon grew exceedingly tame; would allow himself to be scratched upon the head with a straw; would take food from my hand, and many were the games he joined me in, racing about the tank, rising to leap at my fingers, and darting around with every evidence of enjoyment. In feeding he would often jump out of the water repeatedly, as far as the pectoral fins, in trying to take the morsel, in many ways convincing me that he was a very remarkable fish.

Kindness with animals rarely fails, if I except the case of a favorite fox-hound Vän, who became so learned that she recognized the proud cackle of the hens when an egg was laid, and took occasion to rush down to the nests and eat them while they were warm and fresh. Words were thrown away upon her, and even while receiving a salutary whipping, her great brown eyes had a look in them as though she knew it was all humbug, and was being done merely for appearance.

CHAPTER IX.

IS THERE A SEA-SERPENT?

WHEN Sir Charles Lyell asked Colonel T. H. Perkins, a well-known resident of Boston, whether he had heard of the sea-serpent, the reply was, "Unfortunately, I have seen it," and although this answer was given many years ago, it would serve equally well for the present day, as all those who have seen the great unknown are looked upon with a suspicious eye and innumerable arguments brought to bear to prove that they are mistaken. The time has come, however, when the statements of well-known men should at least meet with respectful consideration. For many years the reports of the gigantic squid were considered fabulous, and it is within ten years only that these animals, possessing a length of fifty feet or so, have been accepted by an exacting public. If these large creatures have so long remained concealed, the deep sea may easily hide even larger forms that might readily correspond with the sea-serpent described in current literature. One of the commonest sights along shore is the porpoise and dolphin, yet rarely are these animals found dead or their remains washed ashore, and if they and many others are destroyed by natural causes, how possible it would be for any larger animal to remain undiscovered.

To show plainly the evidence in favor of the existence of the sea-serpent, the following testimony is given; a letter from an old friend, followed by a description of the actual sea-serpents whose remains are found in the geological horizons of the ancient world.

91

"Lynn, Mass., June 26th, 1881.

"Mr. C. F. Holder:

"*Dear Sir:* Yours of the 24th inst. came duly to hand, and, in reply to that part of it relating to the account given by myself of a strange fish, serpent, or some other marine animal called a sea-serpent, I have to say that I saw him on a pleasant, calm summer morning of August, 1819, from Long Beach, Lynn, now Nahant.

"At this time he was about a quarter of a mile away; but the water was so smooth that I could plainly see his head and the motion of his body, but not distinctly enough to give a good description of him. Later in the day I saw him again off 'Red Rock.' He then passed along one hundred feet from where I stood, with head about two feet out of the water, and his speed was about the ordinary of a common steamer. What I saw of his length was from fifty to sixty feet.

"It was very difficult to count the bunches, or humps (not fins), upon his back, as by the undulating motion they did not all appear at once. This accounts in part for the varied descriptions given of him by different parties. His appearance on the surface of the water was occasional and but for a short time. The color of his skin was dark, differing but little from the water or the back of any common fish. This is the best description I can give of him from my own observation. And I saw the monster just as truly, although not quite so clearly, as I ever saw anything.

"This matter has been treated by many as a hoax, fish-story, or a sea-side phenomenon, to bring trade and profit to the watering-places; but, notwithstanding all this, there is no doubt in my mind that some kind of an uncommon and strange rover in the form of a snake or serpent, called an ichthyosaurus, plesiosaurus, or some other long-named marine animal, has been seen by hundreds of men and boys in our own, if not in other waters. And five persons besides myself—Amos Lawrence, Samuel Cabot, and James Prince, of Boston, Benjamin F. Newhall of Saugus, and John Marston of Swampscott—bore public testimony of seeing him at the time.   Yours truly,

"NATHAN D. CHASE."

This appearance attracted so much attention that the Boston Linnæan Society—the scientific society of the time—sent a committee to report upon it.   Dr. Bigelow and Mr. F. C. Gray were selected, and drew up a report signed by a number of witnesses who were within fair sight of the creature.   "The monster," they say, "was from eighty to ninety feet long, his head usually carried about two feet above water; of a dark brown color; the body with thirty or more protuberances, compared by some to four gallon kegs, by others to a string of buoys, and called by several persons bunches, on the back; motions very rapid, faster than those of a whale, swimming a mile in three minutes, and sometimes more, leaving a wake behind him; chasing mackerel, herrings, and other fish, which were seen jumping out of the water, many at a time, as he approached. He only came to the surface of the sea in calm and bright weather.   A skillful gunner fired at him from a boat, and, having taken good aim, felt sure he must have hit him on the head; the creature turned toward him, then dived under the boat, and reappeared one hundred yards on the other side."

Mr. Amos Lawrence writes of the same animal: "I have never had any doubt of the existence of the sea-serpent since the morning he was seen off Nahant by old Marshal Prince through his famous spy-glass. For within the next two hours I conversed with Mr. Samuel Cabot and Mr. Daniel P. Parker, I think, who had spent a part of that morning in witnessing its movements. In addition, Colonel Harris, the commander at Fort Independence, told me that the creature had been seen by a number of his soldiers while standing sentry in the early dawn, some time before this show at Nahant; and Colonel Harris as firmly believed it as though the creature were drawn up before us in State Street, where we then were. I again say, I have never, from that day, to this, had a doubt of the sea-serpent's existence."

James Prince, Esq., then marshal of the district, writes as follows to the Hon. Judge Davis:

"*My dear Judge:* I presume I may have seen what is generally thought to be the sea-serpent. I have also seen my name in the evening newspaper printed at Boston, on Saturday in a communication on this subject. For your gratification and from a desire that my name may not sanction anything beyond what was actually presented and passed in review before me, I will now state that which in the presence of more than two hundred other witnesses took place near the Long Beach of Nahant on Saturday morning last.

"Intending to pass two or three days with my family at Nahant, we left Boston early on Saturday morning. On passing the Half-Way House, on the Salem Turnpike, Mr. Smith informed us that the sea-serpent had been seen the evening before at Nahant beach, and that a vast number of people from Lynn had gone to the beach that morning in hopes of being gratified with a sight of him. This was confirmed at the hotel. I was glad to find that I had brought my famous mast-head spy-glass with me, as it would enable me, from its form and size,

to view him to advantage if I might be so fortunate as to see
him. On our arrival on the beach, we associated with a considerable collection of persons on foot and chaises; and very soon
an animal of the fish kind made his appearance. . . . His
head appeared about three feet out of water; I counted thirteen
bunches on his back; my family thought there were fifteen. He
passed three times at a moderate rate across the bay, but so fleet
as to occasion a foam in the water; and my family and self, who
were in a carriage, judged that he was from fifty to not more
than sixty feet in length. Whether, however, the wake might
not add to the appearance of his length, or whether the undulations of the water or his peculiar manner of propelling himself
might not cause the appearance of protuberances, I leave for
your better judgment. The first view of the animal caused some
agitation, and the novelty perhaps prevented that precise dis-
crimination which afterward took place. As he swam up the
bay, we and the other spectators moved on and kept nearly
abreast of him. He occasionally withdrew himself under water,
and the idea occurred to me that his occasionally raising his
head above the level of the water was to take breath, as the time
he kept under was, on an average, about eight minutes.

. . . "Mrs. Prince and the coachman, having better eyes
than myself, were of great assistance to me in marking the
progress of the animal. They would say, 'He is now turning;'
and by the aid of my glass I distinctly saw him in this movement. He did not turn without occupying some space, and
taking into view the time and the space which he found necessary for his ease and accommodation, I adopted it as a criterion to
form some judgment of his length. I had seven distinct views
of him from the Long Beach, so called, and at some of them the
animal was not more than a hundred yards distant. After we

had been on the Long Beach with other spectators about an hour, the animal disappeared, and I proceeded on toward Nahant; but on passing the second beach I met Mr. James Magee, of Boston, with several ladies in a carriage, prompted by curiosity to endeavor to see the animal; and we were again gratified beyond even what we saw in the other bay, which I concluded he had left in consequence of the number of boats in the offing in pursuit of him, the noise of whose oars must have disturbed him, as he appeared to us to be a harmless timid animal. We had here more than a dozen different views of him, and each similar to the other,—one, however, so near that the coachman exclaimed, 'Oh, see his glistening eye!' Certain it is, he is a very strange animal."

Among the papers of the late Benjamin F. Newhall, of Saugus, is an interesting account of what he witnessed of the seeming gambols of the monster, who appeared to him also to be a timid animal. " As he approached the shore, about nine A. M.," says Mr. Newhall, " he raised his head apparently about six feet, and moved very rapidly. I could see the white spray each side of his neck as he plunged through the water." He came so near as to startle many of the spectators, and then suddenly retreated. "As he turned short, the snake-like form became apparent, bending like an eel. I could see plainly what appeared a succession of bunches, or humps, upon his back, which the sun caused to glisten like glass."

As most of these observers were not seafaring men, their evidence might be doubted from their not being perfectly familiar with marine animals. To show, however, that all classes agreed upon the main particulars, I give the following: " John Marston, a respectable and credible resident of Swampscott, appeared before a justice of the peace and made oath that

PLATE XII.

SWORDFISH ATTACKING A DORY.

as he was walking over Nahant Beach, on the 3d of August, his attention was suddenly arrested by seeing in the water, within two or three hundred yards of the shore, a singular-looking fish in the form of a serpent. He had a fair view of him, and at once concluded that he was the veritable sea-serpent. His head was out of water to the extent of about a foot, and he remained in view from fifteen to twenty minutes, when he swam off toward King's Beach. Mr. Marston judged that the animal was from eighty to a hundred feet in length, and he says : ' I saw the whole body of the serpent—not his wake, but the fish itself. It would rise in the water with an undulatory motion, and then all his body would sink, except his head. Then his body would rise again. His head was above water all the time. This was about eight o'clock A. M. It was quite calm. I have been constantly engaged in fishing since my youth, and I have seen all sorts of fishes and hundreds of horse-mackerel, but I never before saw anything like this.' "

A further example of what might be called expert testimony, is furnished in that of the crew of the bark " Pauline," of London. Their testimony was taken before the stipendiary magistrate at the Liverpool court :

" Borough of Liverpool, in the county Palatine of Lancaster, to wit : We, the undersigned, captain, officers, and crew of the bark ' Pauline,' (of London), of Liverpool, in the county of Lancaster, in the United Kingdom of Great Britain and Ireland, do solemnly and sincerely declare that on July 8th, 1875, in latitude 5° 13′ s., longitude 35° w., we observed three large sperm-whales, and one of them was gripped round the body with two turns of what appeared to be a huge serpent. The head and tail appeared to have a length beyond the coils of about thirty feet, and its girth eight or nine feet. The serpent whirled its

7

victim round and round for about fifteen minutes, and then suddenly dragged the whale to the bottom, head first. George Drevar, *master;* Horatio Thompson, John Henderson Landells, Owen Baker, William Lewarn.

"Again, on July 13th, a similar serpent was seen about two hundred yards off, shooting itself along the surface, the head and neck being out of the water several feet. This was seen only by the captain and one ordinary seaman, whose signatures are affixed: George Drevar, *master;* Owen Baker.

"A few moments after, it was seen elevated some sixty feet perpendicularly in the air by the chief officer and the following able seamen, whose signatures are also affixed: Horatio Thompson, William Lewarn, Owen Baker."

The well-known geologist J. W. Dawson, states that a sea-monster appeared at Marigomish, in the gulf of St. Lawrence, about one hundred feet long, and was seen by two intelligent observers, nearly aground in calm water, within two hundred feet of the beach, where it remained in sight about half an hour and then got off with difficulty. One of the witnesses went up a bank, in order to look down upon it. They said it sometimes raised its head (which resembled that of a seal) partly out of the water. Along its back were a number of humps or protuberances, which, in the opinion of the observer on the beach, were true humps, while the other thought they were produced by vertical flexures of the body. Between the head and the first protuberance there was a straight part of the back of considerable length, and this part was generally above water. The color appeared black, and the skin had a rough appearance. The animal was seen to bend its body almost into a circle and again to unbend it with great rapidity. It was slender in proportion to its length. After it had disappeared in deep water, its wake was visible for some time. Some other persons who saw it compared the creature to a long string of fishing-net buoys moving rapidly about. In the course of the summer, the fishermen on the eastern shore of Prince Edward's Island, in the Gulf of St. Lawrence, had been terrified by this sea-monster; and the year before, a similar creature swam slowly past the pier at Arisaig, near the east end of Nova Scotia, and, there being only a slight breeze at the time, was attentively observed by Mr. Barry, a mill-wright, of Pictou, who told Mr. Dawson he was within one hundred and twenty feet of it, and estimated its length at sixty feet, and the thickness of its body at three feet. It had humps on the back, which seemed too small and close together to be bends of the body. The body appeared also to move in *long undulations*, including many of the smaller humps. In consequence of this motion the head and tail were sometimes both out of sight and sometimes both above water. The head was rounded and obtuse in front, and was never elevated more than a foot above the surface. The tail was pointed, appearing like half a mackerel's tail. The color of the part seen was black. It was suggested by Mr. Dawson that a swell in the sea might give the deceptive appearance of an undulating movement, as it is well known "that a stick held horizontally at the surface of water when there is a ripple seems to have an uneven outline." But Mr. Barry replied that he observed the animal very attentively, having read accounts of the sea-serpent, and felt confident that the undulations were not those of the water.

Professor Richard A. Proctor, the well-known astronomer, gives the following account of a remarkable sea animal, which, however, I am inclined to think was a form allied to the giant-squids, I have often observed squids in Southern waters, especially when pursued, rushing along at the surface, the arrow-shaped caudal extremity elevated above the water and coming down at intervals with a splash, looking, in truth, like the head of a snake, while the tentacles dragging behind formed ripples and convolutions that might easily have suggested the motions of an animal of serpentine form. Mr. Proctor says, "Soon after the British steamship 'Nestor' anchored at Shanghai, last October, John K. Webster, the captain, and James Anderson, the ship's surgeon, appeared before the acting law-secretary in the British Supreme Court, and made affidavit to the following effect: On September 11th, at half-past ten A. M., fifteen miles northwest of North Sand lighthouse, in the Malacca Straits, the weather being fine and the sea smooth, the captain saw an object which had been pointed out by the third officer as a 'shoal.' 'Surprised at finding a shoal in such a well-known tract, I watched the object and found that it was in motion, keeping up the same speed with the ship, and retaining about the same distance as first seen. The shape of the creature I would compare to that of a gigantic frog. The head, of a pale yellowish color, was about twenty feet in length and six feet of the crown were above the water. I tried in vain to make out the eyes and mouth: the mouth, however, may have been below water. The head was immediately connected with the body, without any indication of a neck. The body was about forty-five or fifty feet long, and of an oval shape, perfectly smooth, but there may have been a slight ridge along the spine. The back rose some five feet above the surface. An immense tail, fully one hundred and fifty feet in length, rose a few inches above the water. This tail I saw distinctly from its junction with the body to its extremity: it seemed cylindrical, with a very slight taper, and I estimated its diameter at four feet. The body and tail were marked with alternate bands of stripes, black and pale yellow in color. The stripes were distinct to the very end of the tail. I cannot say whether the tail terminated in a fin or not. The creature possessed no fins or paddles, so far as we could perceive. I cannot say if it had legs. It appeared to progress by means of an undulatory motion of the tail in a vertical plane (that is, up and down).'

"Mr. Anderson, the surgeon, confirmed the captain's account in all essential respects. He regarded the creature as an enormous marine salamander. It was apparently of a gelatinous (that is, flabby) substance. Though keeping up with us, at the rate of nearly ten knots an hour, its movements seemed lethargic. I saw no eyes or fins, and am certain that the creature did not blow or spout in the manner of a whale. I should not compare it for a moment to a snake. The only creatures it could be compared with are the newt or frog tribe." Probably a squid.

The above accounts give a general idea of the appearance of the supposed sea-serpents, and now let us examine some of the fishes known to science, and see if any of them could possibly be taken for one of these forms. The band or tape-fishes from their snake-like appearance, are first worthy of notice, one, the *Regalecus Banksii*, attaining a remarkable length. The largest known was captured by one of Lord Norbury's smacks in the Frith of Forth, Scotland, and was about sixty feet in length, eight or nine inches in width, and altogether a wonderfully slender creature. If swimming at the surface, it might have looked very much like a snake. These ribbon fishes are deep water forms, rarely coming to the surface, and are found in various seas.

THE PEMAQUID SEA-SERPENT.

Quite recently a remarkable fish has been brought to the attention of the director of the National Museum. It was twenty-five feet long, about eight inches wide, and called a sea-serpent by the fishermen who discovered it. It was caught off Pemaquid Point, Maine, and the following extracts of letters were written by the fisherman explaining his find, the accompanying illustration (*Plate XVI.*) having been made from his sketch that was forwarded to Professor Baird. "The fish was about twenty-five feet in length and from eight to ten inches in diameter, with a tail like an eel. The skin was not like a scale-fish, but more like a dog-fish or shark, though a great deal finer in quality. I did not save the fish for the reason that I did not know what I had caught. In fact I considered it a streak of ill-luck rather than good fortune, having torn my nets very badly and otherwise bothering me in my business. The fish could have been grappled twenty-four hours after, it being in only four fathoms of water and it being a small shoal, with deep water all around it. A storm arose later, which made it impossible to do so. . . . Exclusive of the head, it looked very much like an eel. The body was round or very near that form. The tail was like that of a common eel. The color of its back was of a slate or fish color; belly, grayish-white. There were two fins, one on either side, a little abaft the head. They were not stiff-pointed fins like the shark, or sword-fish, but more like the side fins of the cod or sun-fish, only they were in size to correspond with the fish. The top or dorsal fin was like the corresponding fin on the cod. I do not know whether it was stationary or closed, like the top fin of the mackerel and other fish of the same species. All the fins that were on the tail were like that of the eel. The skin was like that of the dog-fish, only very much finer. The head resembled that of the shark, though more stunted, *i. e.*, it did not lengthen out like the shark's. It looked more like the head of the sucker. The mouth was very small, not any larger than that of a good-sized dog-fish, with fine briery teeth, and located at the extreme end of the head or nose. The fish was dead when caught."

This fish is possibly a form similar to the new Japanese eel-like shark *Chlamydoselachus anguineus*, recently described by Mr. Garman. Regarding it he says : " Such an animal is likely to unsettle disbelief in what is popularly called the ' sea serpent.' In view of the possible discoveries of the future, the fact of the existence' of such creatures, so recently undiscovered, certainly calls for a suspension of judgment in regard to the non-existence of that oft-appearing but elusive creature, the serpent-like monster of the oceans."

It has been suggested that several sharks or porpoises swimming one behind the other would correspond in appearance with the descriptions given of the sea-serpent ; yet it seems impossible that so many persons familiar with the sea should have been deceived.

It has also been claimed that if the sea-serpent exists that it is a descendant of some of the wonderful serpentine creatures that lived in former geological ages, and a glance at the results of the investigations of Profs. Leidy, Cope, and others, shows that even if this is not so, sea-serpents in all the term implies, were the common features of these past ages.

The Mesozoic time, or the Age of Reptiles, in which the ancient sea-serpents lived, includes, according to Cope, the Triassic, Jurassic, and Cretaceous periods ; and the period of transition from the Palaeozoic to it is strongly marked. A great change was impending, and a nearly complete extermination of existing life took place. In the new era came the monster forms with which we have become familiar within a few years. During the Cretaceous, in the limestones of which in Kansas and New Jersey are found some of the most interesting creatures, the North American continent presented a strange contrast to its present state. Florida was not yet above water, nor any of the border States, while a great sea extended from the gulf of Mexico northwest. The old coast-line can be readily traced, and extended from Arkansas to near Fort Riley, on the Kansas River, passing to the east through Minnesota to Canada, near the head of Lake Superior, while to the west it spread away to an unknown distance, the shore probably now submerged by the Pacific. Such was the Cretaceous sea, and now cities, towns, and railroads are dotted over the region, while immense desert tracts mark other portions, where water is now never seen. But only yesterday in the age of the world another scene was being enacted, which is further described by Professor Cope in substance in the following :

SEA-SERPENTS OF SCIENCE.

Far out on the expanse of this ancient sea might have been seen a huge snake-like form, which rose above the surface and stood erect, with tapering throat and arrow-shaped head, or swayed about, describing a circle of twenty feet radius above the water ; then, plunging into the depths, naught would be visible but the foam caused by the disappearing mass of life. Should several have appeared together, we can easily imagine tall, flexible forms rising to the height of the masts of a fishing-fleet, or, like snakes, twisting and knotting themselves together. This extraordinary neck—for such it was—rose from a body of elephantine proportions, and a tail of the serpent pattern balanced it behind. The limbs were probably two pairs of paddles, like those of the *Plesiosaurus*, from which this diver chiefly differed in the arrangement of the bones of the chest. In the best-known species twenty-two feet represent the neck in a total length of fifty feet. This

PLATE XIII

THE BRILLIANT LAMP FISH.

is the *Elasmosaurus platyurus* a carnivorous sea-reptile (*Plate XVII.*), no doubt adapted for deeper waters than many of the others. Like the snake-bird of Florida, it probably often swam many feet below the surface, raising the head to the distant air for a breath, then withdrawing it, and exploring the depths forty feet below without altering the position of its body. The general form of this reptile, a fine skeleton of which can be seen in the museum of the Academy of Sciences, Philadelphia, was that of a serpent, with a relatively shorter, more robust, and more posteriorly placed body than is characteristic of true serpents, and with two pairs of limbs, or paddles. It progressed by the strokes of its paddles, assisted by its powerful and oar-like tail. The snake-like neck was raised high in air or depressed at the will of the animal, now arched swan-like preparatory to a plunge after a fish, now stretched in repose on the water, or deflexed in exploring the depths below. Researches into their structure have shown that these creatures were of wonderful elongation of form, especially of tail ; that their heads were large, flat, and conical, with eyes directed partly upward; that they were furnished with two pairs of paddles, like the flippers of a whale, attached by short, wide peduncles to the body. With these flippers and the eel-like strokes of their flattened tail they swam, some with less, others with greater speed. They were furnished like snakes with four rows of formidable teeth on the roof of the mouth. Though these were not designed for mastication, and in the absence of paws for grasping, could have been little used for cutting, as weapons for seizing their prey they were very formidable.

These sea-serpents swallowed their prey entire, being able to do so by a peculiar arrangement of the jaws. It is also assumed by Professor Cope that the only sound they could utter was a serpent-like hiss.

The giants of the *Pythonomorpha* of Kansas have been called *Liodon proriger*, and *Lindon dyspelor*. The first must have been very abundant, and its length could not have been far from seventy-five feet,—certainly not less. Its physiognomy was rendered peculiar by a long, projecting muzzle, reminding one of that of the blunt-nosed sturgeon of our coast; but the resemblance was destroyed by the correspondingly massive end of the branches of the lower jaw. Professor Cope states that he once found the wreck of an individual of this species strewn around a sunny knoll beside a bluff, and its conic snout pointing to the heavens formed a fitting monument, as at once its favorite weapon and the mark distinguishing all its race. The *Lindon dyspelor* was the longest of known reptiles, and probably equal to the great finner-whales of modern oceans.

Another monster snake-like reptile was the Mososaurus, which closely resembles, when restored, the typical sea-serpent of to-day. Professor Marsh, of Yale, says of it: "The reptiles most characteristic of our American cretaceous strata are the *Mososauria*, a group with few representatives in other parts of the world. In our cretaceous seas they ruled supreme, as their numbers, size, and carnivorous habits enabled them to easily vanquish all rivals. Some were at least sixty feet in length, and the smallest, ten or twelve. In the inland cretaceous sea from which the Rocky Mountains were beginning to emerge, these ancient sea-serpents abounded, and many were entombed in its muddy bottom. On one occasion, as I rode through a valley washed out of this old ocean-bed, I saw no less than seven different skeletons of these monsters in sight at once. The Mososaurs were essentially swimming lizards, with four well-developed paddles, and they had little affinity with modern serpents, to which they have been compared." The Clidastes was noted for its elongation, and a specimen, representing an animal from sixty to eighty feet in length, has recently been discovered at Freehold, N. J., by Professor Lockwood, of Rutgers. The teeth were terrible weapons, having fore and aft cutting edges.

Even more remarkable than the above were the *Amphicoelias* and *Camarasaurus* (*Plate XVIII.*), the former attaining a length of one hundred feet, and the latter seventy-five—gigantic serpentine reptiles that floated in shallow waters, anchored by their ponderous tail and legs.

Such were some of the sea-serpents of the Reptilian Age, bones of which the sceptic may find in any of our museums. According to Professor Marsh, the first American serpents, so far as now known, appeared in the Eocene, which contains also the oldest European species.

On the then Atlantic border existed a great sea-snake, at least thirty feet long, known to science as the Titanophis, while about the inland lakes and bodies of water lived large serpentine forms allied to the boa-constrictors of to-day. The true water-snakes of the present time often attain a large size, and might readily pass for sea-serpents. They are known scientifically as the *Hydrophidæ*. Professor Bickmore informed me that on his voyage to the Indian Archipelago, he frequently shot at them in the open ocean. The master of the ship Georgiana claims to have seen one off Rangoon that was fifty feet long, that slowly passed the ship's bows. It was of a gray yellowish hue, and appeared to be about a foot thick. The captain and crew watched it for twenty minutes, examining it well and carefully. These snakes are often venomous, and differ from ordinary land snakes in having a flat paddle-like tail, that enables them to move quickly through the water.

Is the sea-serpent a gigantic form of these snakes, a huge band fish, *Regalecus*, a shark-like ally of *Chlamydoselachus*, or is it a survivor of the age of the *Elasmosaurus?* I have shown that the sea-serpent did exist, and it probably would not surprise paleontologists to find a living form yet alive.* The chances are, however, against such an occurrence, and the sea-serpent possibly may prove a gigantic deep-sea fish that only at occasional intervals appears upon our shores to excite the wonder and amazement of the inhabitants.

* As this work goes to press I have received information from a reliable source that a long, slender animal, forty-two feet by actual measurement, has been found in Southern waters. Its greatest girth was about that of a horse, the tail long and slender. The finder stated that it had bony ribs and a pair of paddles, hence it was not a shark. The small body and attenuated tail do not suggest a whale of forty-two feet. Its length precludes the idea of its being a manatee, and the sketch made by the finder is that of a headless, neckless, Elasmosaurus. The head of the animal was gone; but the finder hauled the body above high water mark, and a party will be sent to determine whether or not it is an animal new to science.

CHAPTER X.

ANIMAL ELECTRICIANS.

How often in wandering by the shore or through some quiet stretch of woodland are we attracted by the ingenious efforts at defence or protection displayed by the lowly creatures that there find homes! Some erect elaborate structures, calculated to deceive by their resemblance to extraneous objects, while many more possess peculiarly aggressive features that furnish effective protection. Among the latter class are a number of fishes that are remarkable electric batteries, presenting a strange resemblance to the electric appliances of human invention.

Nine different fishes, representing several genera, have been found charged by nature in this remarkable manner. Along our Eastern shore, the torpedo—one of the rays, and the best known of the electric animals—is not uncommon, and fishermen frequently find their arms bound in invisible chains and rigid from the message sent up the line from this strange creature.

In the seventeenth century the attention of Redi, the Italian naturalist, was attracted by the tales told by the fishermen, who thought the torpedo was protected by some peculiar witchcraft that overcame them when they attempted its capture. One was brought to the distinguished *savant*, who subjected it to a number of tests. "I had scarcely touched and pressed it with my hand," he writes, "when I experienced a tingling sensation, which extended to my arms and shoulders followed by a disagreeable trembling, with a painful and acute sensation in the elbow-joint

109

that made me withdraw my arm immediately." He also found that these sensations resulting from contact with the fish diminished as the death of the torpedo approached, ceasing altogether as the animal died. Later, Réaumur examined the then problematical subject, and says concerning it, "The benumbing influence is very different from any similar sensation. All over the arm there is a commotion which it is impossible to describe, but which, so far as comparison can be made, resembles the sensation produced by striking the tender part of the elbow against a hard substance."

Neither of these scientists, however, discovered the true nature of the creature's defence,—an honor reserved for Dr. Walsh, of London. During a visit to the Isle of Ré, he and a number of friends amused themselves with these fishes, finally discovering their electrical nature. The battery is constructed on the principle of the voltaic pile, and consists of two layers or series of cells of hexagonal shape, as many as two thousand five hundred being found in a single fish of small size. The space between the numerous delicate transverse plates in the cell is filled with a jelly-like mucous fluid, so that each cell represents to all intents and purposes a Leyden jar. Each cell is provided with nerves, while the dorsal side is positive and the ventral negative. It is supposed that the impression is conveyed by certain nerves to the brain, exciting there an act of the will, which is conveyed along the electric nerves to the batteries producing the shock.

One of the experiments of Dr. Walsh was to place a living torpedo upon a wet cloth or towel; he then suspended from a plate two pieces of brass wire by means of silken cord, which served to isolate them. Round the torpedo were eight persons standing on isolating substances. One end of the brass wire was supported by the wet towel, the other end being placed in a

basin full of water. The first person had a finger of one hand
in this basin, and the finger of the other in a second basin, also
full of water. The second person placed a finger of one hand in
this second basin and a finger of the other hand in a third basin.
The third person did the same; and so on, until a perfect chain
was established between the eight persons and the nine basins.
Into the ninth basin the end of the second brass wire was
plunged, while Dr. Walsh applied the other end to the back of
the torpedo, thus establishing a complete conducting circle. At
the moment when the experimenter touched the torpedo, the
eight actors in the experiment felt a sudden shock, similar in all
respects to that communicated by the shock of a Leyden jar,
only less intense. When the torpedo was placed on an isolated
supporter, it communicated to many persons similarly placed
from forty to fifty shocks in a minute and a half. Each effort
made by the animal in order to give them was accompanied by
the depression of its eyes, which seemed to be drawn within their
orbits, while the other parts of the body remained immovable.
If but one of the two organs of the torpedo were touched, only
a slight sensation was experienced,—a numbness rather than a
shock. When the animal was tried with a non-conducting
rod, no shock followed; glass, or a rod covered with wax, produced no effect; touched with a metallic wire, a violent shock
followed. Melloni, Matteucci, Becquerel, and Breschet have all
made the same experiment, with the same results,—Matteucci
having ascertained that the shock produced by the torpedo is
comparable to that given by a voltaic pile of a hundred to two
hundred and fifty pairs of plates.

The experiments of Dr. Walsh produced an electric craze in
England, and the demand for torpedoes was unprecedented.
Their curative powers were extolled, and large sums were paid

by invalids for opportunities to test their effects. On old Brighton Beach a large torpedo or cramp-fish was exhibited in a shallow water aquarium by an enterprising showman, who proclaimed to the assembled multitudes that he had on exhibition "the heaviest fish in the world,—heavier than a whale, and brought in a single ship all the way from the Antarctic Ocean!" He furthermore stated that a ha'penny would be accepted as a consideration for the privilege of lifting the fish, and a shilling would be given to any one who should lift it out of the tank bare-handed. This enticing offer was taken by numbers of muscular sojourners on the beach, but always resulted disastrously to the lifter, who, however, was unable to explain why he had failed. Another would step boldly up with bared arms, insert one hand carefully under the fish to see that it was not held down (just what the showman wished him to do), and place the other upon the torpedo's back. Its queer eyes would wink, a convulsive movement followed, and the experimenter would find himself either unable to move or almost lifted into the air by the "heft" of the creature, and would fall back bewildered, amid the jeers and laughter of the crowd.

The effect of the shock upon birds is generally fatal. A reed-bird placed in the water over a torpedo showed symptoms of fear almost immediately, and in less than two minutes dropped dead. Although the torpedo does not heed its own shocks, and is used as an article of food on the Mediterranean coast, it is particularly sensible to shocks administered by a regular battery, and can thus be readily killed. Its power is hardly sufficient to kill a man, though I have been told by a reliable informant that he was almost completely paralyzed when spearing one, and on attempting to pull the iron from the fish he was knocked over as suddenly as if shot. Even after the death of the

PLATE XIV.

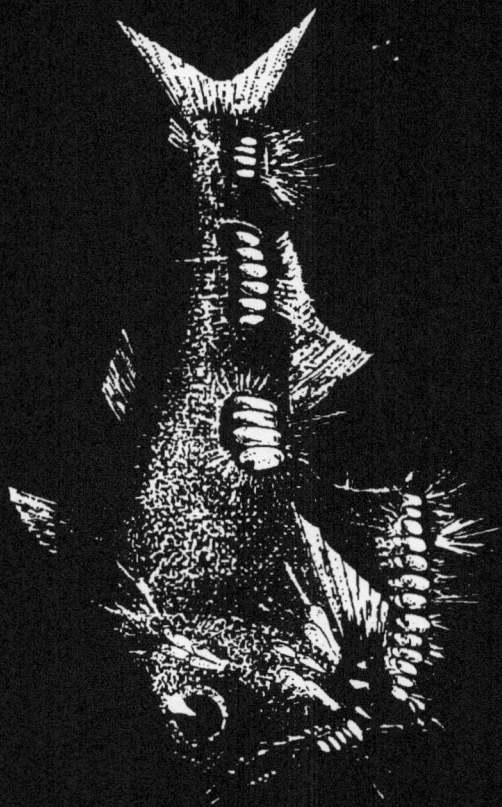

THE LUMINOUS ARGYROPELETUS.

torpedo he could hardly hold the dissecting knife, so intense were the shocks.

In 1671 the astronomer Richer visited Cayenne as a representative of the Paris Academy of Sciences on the geodesic survey. During a fishing trip on one of the streams of the neighborhood, he made an involuntary experiment which few would care to repeat. Having hooked a large fish, he found that his arms were powerless, and the whole upper portion of his body became rigid, as if paralyzed. The natives detached the line from his hand, and for half an hour he remained overcome by the strange attack. Later he was informed by the natives that he had been bewitched by an eel (the gymnotus) which inhabited those waters and frequently killed animals by merely touching them. Richer's experience was detailed to the French Academy, but the *savants* were perhaps incredulous, and the matter was forgotten until seventy years later, when Condamine, the naturalist, visited South America and revived it. Later, in 1755, an eminent Dutch surgeon, Gramund, found that "the effect produced by the fish corresponded exactly with that produced by the Leyden jar, with this difference, that we see no tinsel on its body, however strong the blow it gives, for, if the fish is large, those who touch it are struck down and feel the blow on their whole body."

Humboldt also examined the gymnotus, and gradually the power of this remarkable living battery became generally known. One was quite recently captured near Calabozo, which not only killed a mule, but so prostrated the rider by its terrible powers that his life was despaired of. An English traveler reached the spot a few days after the occurrence, and, learning the size of the monster, determined to catch it. It was finally hooked and dragged upon the shore. The line, however, becoming wet, the fish communicated to the two natives who were holding it

8

such a shock that they were utterly powerless to move. The Englishman rushed forward, cut the rope with a knife, and released the men, but received a shock himself. The fish was finally secured, and a load of shot sent into its head. The men then took hold of its tail to drag it to the bank above, when they were knocked over as if by an axe, and nothing could induce them to touch it again. Not till three days after, when decomposition had probably set in, was it dragged from the shore and suspended from a tree, and skinned with the intention of sending the dried skin to the British Museum, where it would have been placed, but for the ants, who succeeded, in less than a month, in reducing it to tissue.

These gigantic eel-like creatures are most forbidding in appearance, varying from six to twenty-two feet in length, having the same relative size throughout their entire length. The head is broad, the tail compressed, and along its under surface lie the four batteries, two on each side, the mass occupying nearly the whole lower half of the trunk. The curious plates are vertical, instead of horizontal, as in the torpedo, and the entire batteries or cells are horizontal instead of vertical, as in the same fish, each being supplied with nerves by the ventral branches of nearly four hundred spinal nerves. With such an armament they are to be dreaded indeed. A touch of their long bodies is death to fish larger than themselves.

In the streams about Caraccas, South America, are famous spots for these much dreaded fishes, while so common are they in a small lake near Calabozo, that they are caught by thousands. This is done by a singular method, called *embarbascar con caballos*, or intoxicating by means of horses. Mules, horses, and other animals are used, and the scene, though frightfully cruel, is made the occasion of great festivities. The poor

animals are driven by shouts and blows into the water, where
they dash about as if aware of their danger. Great eel-like,
yellow bodies appear, their backs flashing in the sun, darting
about, hurling themselves against the terrified beasts, which with
staring eyes and trembling frames are completely paralyzed by
the electric discharges. Some are killed as if by lightning, and
fall among the writhing mass; others endeavor to break through
the howling throng of natives upon the banks, but are beaten
back to terrible death or torture. The eels seem to be aware of
the most vulnerable points of attack, as they strike the poor
brutes near the heart, discharging the whole length of their
battery. The terrible struggles last from twenty to thirty
minutes, and then those horses that have survived the ordeal
seem to grow careless of the attacks. The fishes have exhausted
their electric supply for the time; and now the natives step to
the fore. The eels, finding their power on the wane, seek the
bottom of the lake; the natives, mounting the horses, rush
wildly about among the fleeing animals, striking them with
their long spears, and dragging them ashore, or anon rolling
from their horses, paralyzed by unexpected shocks that dart up
the wet lines. Great numbers of eels are captured, and it is
always found that, though they soon exhaust their force, if an
attack is intended the next day the same precautions are necessary, their recovery of vital force being extremely rapid.

In 1842 two of these creatures were carried to London, and
kept alive for six years, during which time they doubled their
weight each year. They were examined and experimented upon
by most of the scientific men of the day, and considered remarkable curiosities. "I was so fortunate," says Professor Owen,
"as to witness the experiments performed by Professor Faraday
on the large gymnotus which was so long preserved alive at the

Adelaide Gallery in London. That the most powerful shocks were received when one hand grasped the head and the other hand the tail of the gymnotus I had painful experience, especially at the wrists, the elbow, and across the back. But our distinguished experimenter showed us that the nearer the hands were together, within certain limits, the less powerful was the shock. He demonstrated by the galvanometer that the direction of the electric current was always from the anterior parts of the animal to the posterior parts, and that the person touching the fish with both hands received only the discharge of the parts of the organs included between the points of contact. Needles were converted into magnets, iodine was obtained by polar decomposition of iodide of potassium; and, availing himself of this test, Professor Faraday showed that any given part of the organ is negative to other parts before it, and positive to such as are behind it. Finally heat was evolved, and the electric spark obtained."

There is a story current that a few years ago a firm in Boston ordered a number of gymnotes from their agent in Rio. The fish were duly shipped on a fruit schooner, which was forced by rough weather to make the Bermuda Islands. During a stay there of several days, the crew were continually annoyed by numbers of colored visitors who insisted upon coming aboard, sampling the cargo with such pertinacity that its entire depletion was threatened. They seemed possessed with the demon of curiosity. One huge black was especially obtrusive; nothing was sacred. He went aloft, scoured the hold, examined the galley, and finally lifted the tin cover of the can containing the gymnotes.

"What's dis yer?" he asked the skipper. The latter, who was sitting on the rail, meditatively rubbed his nose, and winking at the cook, replied, "Them's Fiji eels; we swapped off the first mate for 'em out in the Cannibal Islands."

"Is dey big?" questioned the astonished darkey, whirling the water about, and endeavoring to make out the fish.

"They're jest so big," returned the skipper, knocking the ashes out of his pipe, "that ef you'll lift one out on to the deck, I'll give you the best bunch of bananas on the Ann Eliza."

"Dat settles it," rejoined the darkey. "Why, Captin, I'se de boss eeler on dis yer reef; ketch murries all de times, two or tree feet long."

"Waal, they hain't Fiji eels," retorted the mariner. "You don't look to me as ef you hed the necessary muscle."

The native was a brawny specimen, weighing at least two hundred pounds, and this last speech was too much for him. Motioning back some of his companions who had joined him, the "boss eeler" reached into the can, and, cautiously moving about, secured a hold with one hand, while he made a quick grasp with the other and straightened up. A howl that might have been heard half a mile broke from him as he rose up with an enormous eel writhing in his rigid arms. His eyes fairly stood out while he roared and cried in what was veritable anguish.

"What's de matter?" shouted a comrade; "de fish ain't a-bitin' yo?"

"Take it off!" cried the champion.

Thus appealed to, the other seized the eel, and, being a weaker party, was knocked fairly over. Completely demoralized, the entire company, headed by the two victims, now made for the shore, averring that they had been "voodooed," and nearly killed by the Yankee captain's fish.

It would indeed be remarkable did we not find an electrician among the siluroids or cat-fishes which abound in peculiar characteristics. The *Malapterus electricus* of the Nile is one. The electric cells form a layer directly beneath the skin, and envelop the entire body except the head and fins, the creature finding in it perfect protection. The cells are, however, extremely small, about one and a half millimetres in diameter,—and lozenge-shaped. Several specimens of these cat-fishes were exhibited in New York in 1876, and were found to impart a decided shock, though not comparable to that of the torpedo or gymnote. The natives in North Africa call them *raad* or thunder,—certainly a suggestive title; while on the Niger the native name is *Ishenza*.

An electric balloon fish—*Tetraodon*—has been discovered in the waters about the Comoro Islands. These curious fishes were found in great numbers among the cavities of the rocks by the crew of an English ship, and when taken from the water they gave sharp and decided shocks, immediately assuming the strange oval shape from which they have derived their English appellation.

Other fishes—nine in all—are known to be electricians of more or less power, but as yet little is known of their natural method of using their curious defence. That it is such is hardly to be doubted; yet the torpedo is infested with a parasite that bores into its various parts, utterly insensible to the batteries of its victim. Professor Leydig, the eminent Swiss naturalist, marshals the forces of a small army of believers in the electric properties of the mother of pearl spots found in the luminous *Chauliodus*, and other fishes indigenous to the Mediterranean waters. The alleged electric organs are oval spots, generally scattered over the ventral surface, which, when critically examined, appear to bear a resemblance to the electric or pseudo-electric organs of other fishes.

Of all the electric animals, the insects are perhaps the most interesting, possibly from the fact that but little is yet known concerning them. The late General Davis, of the British Army, was the first to discover these insect batteries. His experiments were chiefly confined to a wheel-bug (*Reduvius*) of the West India Islands. In picking one up from the ground he received a decided shock, as if from an electric jar, which affected his arms as high as the elbow. Shaking the insect off, he observed six marks where its feet had been, and from this he inferred that the legs were the electric organs. Other instances of insect electricians have been communicated to the London Entomological Society by Mr. Farrel. One is referred to in a letter from Lady de Gray, of Groby, in which the shock was caused by one of the beetles (*Elateridae*),—so powerful that the arm of the experimenter was rendered useless for some moments. Captain Blakeney, R. N. had a most remarkable experience in South America. Observing a large hairy lepidopterous caterpillar, he attempted to pick it up, when he experienced so powerful an electric shock that his right arm and side were almost paralyzed: his life was, in fact, considered in danger, the force of the discharge being as powerful as that of the torpedo, and more subtile.

CHAPTER XI.

OUR COMMON SNAKES.

PERHAPS no class of animals so widely distributed and so common in every-day life are so little known and understood as the snakes. This is not because the study has been neglected or overlooked, as the scientific institutions of the various cities of the country are replete with fine collections of most of the reptiles and exhaustive works upon their habits and customs. Yet, notwithstanding this, the snake still forms the subject of ever-recurring tales, fabulous in the extreme, that seem handed down from generation to generation with a faithfulness that would do credit to the New Mexico Indians who preserve their records in this way. Curiously enough, many of these strange stories are current among those who, from the nature of their life, would be expected to be well and accurately posted upon the habits of the animals. Thus many farmers and horticulturists can be found that religiously believe that the milk-snake (*Ophibolus clericus*) deprives cows and milk-giving animals of their supply of milk. An intelligent farmer informed me that for a long time his cows had failed to give a proper supply of milk, and it was a great mystery until one morning before milking he saw a milk-snake *between two of the cows*, and killed it on the spot, after that having no trouble. This man could not be convinced of his mistake, although he acknowledged that it was impossible that so small a snake could have held a pint of milk, even if forced into its body.

120

PLATE XV

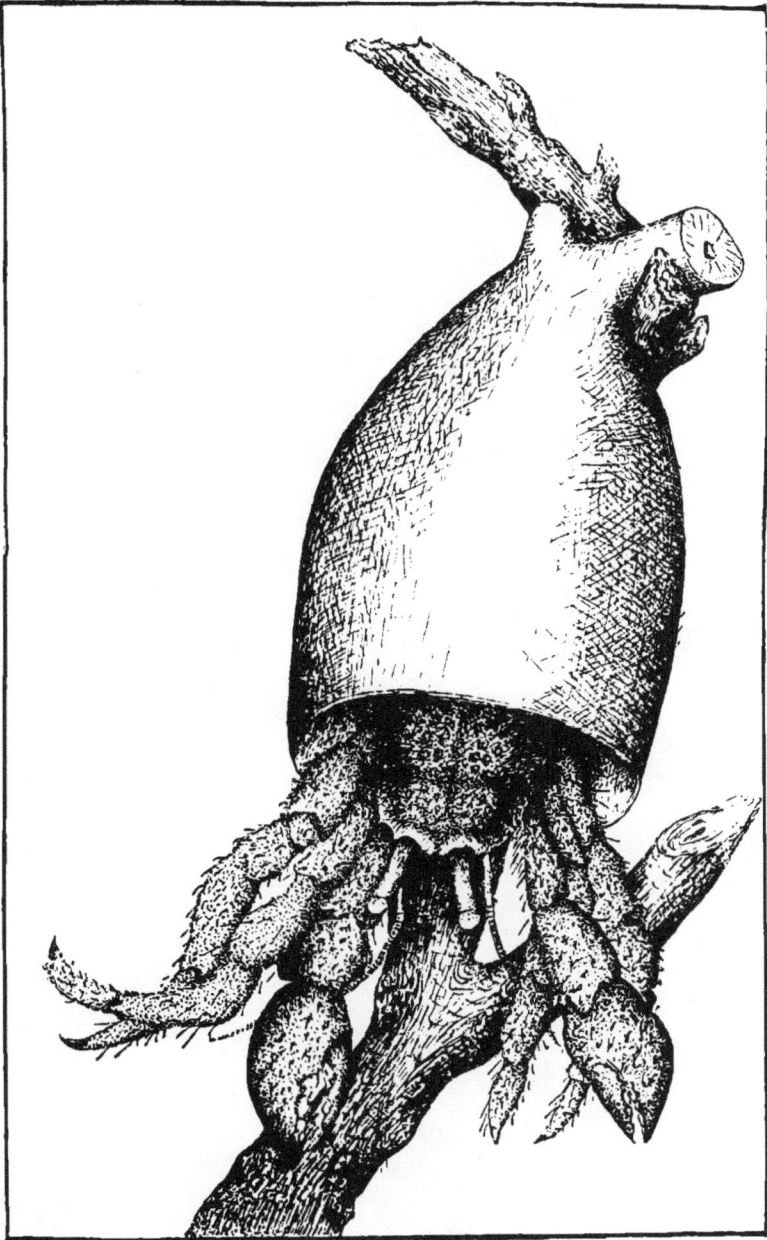

HERMIT CRAB IN A TOBACCO PIPE.

A statement often seen and believed by many, is that a certain snake of the South (the whip-snake) seizes its tail—which has a sting (?)—in its mouth, and rolls away in the form of a wheel, stopping suddenly and striking the enemy with the sting. Such fables are current by the score, and denial only seems to strengthen belief.

We have in America, north of Mexico, about one hundred and thirty species of snakes, nearly all having a wide geographical distribution. They constitute the first order (*Ophidia*) of the reptiles, have long cylindrical bodies, are footless, without a shoulder girdle, and covered with a coat of scales which is shed in the summer months. They are formed from the epidermis, and generally overlap each other as in the fishes; but in other cases, as the *Acrohordus*, etc., they do not, and are tubercular. The eyes of snakes have no lids in the strict sense of the term, being covered permanently by a delicate film, or membrane, that takes the place of the lid, thus explaining the stony stare that all snakes have, and that is so disagreeable to many; so it may be said that the snake cannot shut its eyes.

When we examine the skeleton, we see that it is arranged to allow the greatest amount of freedom and flexibility. The pieces of bone (vertebrae) that go to make up the long tapering backbone number four hundred in some species, are hollow in front and convex behind, literally working on a ball-and-socket plan.

The processes of the vertebrae are provided with what are called articular facets, that lock into or grasp each other, thus strengthening and giving a greater degree of flexibility to the backbone. In the head, however, we see the most remarkable arrangement, that enables the snake to prey upon animals that seem larger than itself. The jawbones would seem to be a combination of elastic springs, having no gauge to their tension;

thus the quadrate bones that connect the lower jaw with the skull are movable, allowing the enormous gape that all who have seen a snake swallow its prey are familiar with. Beside this, the bones of the jaw itself and palate are more or less movable, also tending to assist greatly in the distension of the throat.

As the snakes do not tear or mutilate their prey, their teeth are not set in sockets, and are merely for two purposes: first, to poison and stupefy the prey; and second, to prevent its escape, acting as hooks by which the body is hauled over the victim. We have seen that the bones of the lower jaw were not fastened closely to each other, so in swallowing prey the teeth on one side are advanced, securing a hold, then on the other side, and so on until the victim is hauled into the snake's throat overhand, as if so many hooks were alternately pulling at it.

The poisonous snakes, as the rattlesnakes, have two long sharp fangs, each appearing as if flattened out like a knife blade and bent up, forming a hollow tube, or groove, open at both ends. The upper portion of the hollow fang is fastened to a bone in the cheek which moves with ease, so that when not in use the fangs can be folded—packed away, so to speak—until needed.

Undoubtedly all animals, even man, have in their saliva a deadly poison, though in the latter it is extremely diluted, and of use only in assisting the digestion of our food. In the poisonous snakes, however, we find it accumulated in saes that are modifications of the salivary gland, and placed on each side of the upper jaw. A delicate canal extends from the poison gland under the eye forward to the edge of the jaw, and there opens into the fang above the tube of the tooth; and if we examine here, we shall find rudimentary fangs, all ready to grow out if the large one is lost. To use the poison, the snake has merely to strike its prey, the muscles of the jaw being so arranged that, as soon as the fang enters the flesh of the victim, certain muscles press upon the glands, squeeze the poison through the little canal down through the hollow fang, and the work is done.

In their actions snakes are most graceful. The gliding motion is effected by the movements of the large or ventral scales, that are successively pushed forward, the hinder edges resting on the ground and forming a support. These scales, fulcra, or pushers, are fastened to the ribs by muscles, and by holding a snake the swelling movement can be readily felt.

In color the snakes vary much, but are generally adapted to their surroundings, the green snakes being found in green grass and vegetation, while the grey snakes frequent rocky districts, where they are alike protected. The skin is shed in one piece at various seasons of the year, being forced off in some cases by the snake forming a ring with its tail and squeezing the rest of the body through it, or by wriggling through the bushes. The poisonous snakes may always be recognized by their broad flattened heads, and generally short and thick bodies. They also, as a rule, possess a vertical keel running along the centre of each scale. The non-poisonous snakes have long bodies, with the head small, no distinct neck, and the scales not keeled.

Probably the best known of our common poisonous snakes are the rattlesnakes, that with the moccasins and copperheads form the dangerous family *Crotalidae*, distinguished by the large ugly head and the absence of teeth in the upper jaw, excepting the long fangs, and the pit in the head.

The Northern rattlesnake (*Crotalus horridus*) has probably the widest geographical range, being found in nearly every State in the union, from the borders of the Gulf of Mexico to Northern

New England, and west to the Rocky Mountains. It has a most forbidding appearance, and never could be mistaken when once seen, having an enormous head, triangular in shape, and large brilliant eyes, with a fiery iris. Between the eye and the nostril is a deep pit, that is peculiar to the family.

As the name indicates, they have a rattle that is a horny appendage to the tail, formed of separate button-like objects, that rattle together when the tail is vibrated. This often warns human beings, and perhaps arouses a curiosity in animals that proves fatal. The popular belief that a rattle is added every year is not borne out by facts. In a specimen observed by Dr. Holbrook, two rattles appeared within a year, and Dr. Buchanan's specimens attained four in that period. Mr. Peale, the naturalist, kept one of these snakes fourteen years. When he obtained it it possessed eleven rattles, and though it lost several every year, new ones took their place, and at its death it retained the original number, although, during the fourteen years of confinement in the Philadelphia Museum, it had increased four inches in length. So it will be seen that it is impossible to determine the age of these animals by this means. The number of rattles attained is also uncertain; the greatest number observed by Dr. Holbrook, was twenty-one, all of which were perfect; but a writer in the *Columbian Magazine*, in 1786, recorded a specimen that had forty-four.

The rattlesnake is mild and peaceful if unprovoked, and has never been known to follow a human being for purposes of attack. A friend of the author, an army officer, was sitting on a stump in Florida with his hands upon his knees, looking down upon the ground, when a large rattlesnake, not, however, of this species, slowly crawled between his legs, and moved away. The officer had the nerve to remain perfectly quiet, which was necessary, he said, there being no stimulant within two hundred miles.

Rattlesnakes prey upon small animals, as rats, squirrels and rabbits, and can always be safely approached when stretched out, only striking when they are coiled. They are not climbers, rarely, if ever, being found in trees, and their alleged powers of fascination are purely mythical, this lying merely in the horror their presence inspires in the lower animals as well as in man. Paralyzed by fear, the victim is often incapable of flight, and stupidly awaits its fate. This I think will explain all the so-called cases of fascination, and the reader can well apply it to familiar examples by recalling instances where men, women and children have been attacked by animals, and rooted to the spot, as it were, by fear and surprise. Men have been so horrified and confused at an approaching train, that they have stood until tossed from the track, yet, in entering the plea for damages, the plaintiff's lawyer never claims that the injured party was fascinated by the train.

These dangerous snakes are extremely common in New York State, and especially in the mountainous and secluded parts of Pennsylvania. On the shores of Lake Champlain they are also very abundant, and some years ago two men, in three days, destroyed over eleven hundred rattlesnakes on the sides of Tongue Mountain, in Bolton Township, N. Y.

As one of the vagaries of fashion is to have card-cases, bags, belts, pocket-books, etc., made of the rich skins of these animals, an extensive business is now carried on, and many men are engaged as professional rattlesnake-hunters. Especially in Sullivan and Ulster counties is this curious business followed; and, according to a gentleman living in Middletown, very large sums are made every year in the sale of rattlesnake oil, which is believed to possess wonderful curative powers by a large proportion of the inhabitants of not only those but of adjoining counties. Many snakes are killed during the summer season, but the grand gathering of the crop is in the fall, when they have returned to their dens and wintering-places. These retreats are well known to the snake-hunters, and they choose sunny days in October and November for raiding them. On such days the reptiles crawl out of their dens in the rocks, and huddle together by the score, different varieties frequently being found massed together. The snakes are dull and sluggish at that time of the year, and come out to bask in the sun. The hunters arm themselves with old fashioned flails, and when they come upon a group of snakes, proceed at once to thrash them, but few escaping.

The rattlesnakes are assorted from the other species and carried home, where the oil is simply tried out and bottled up, ready for the market and the credulous patient.

In the winter months these snakes enter upon a state of hibernation, burying themselves in the ground, entwined or singly, coming out in early spring.

DO RATTLESNAKES SWALLOW THEIR YOUNG?

Perhaps no subject connected with snakes has attracted so much attention as the vexed one as to their care for their young. A snake would scarcely be expected to show much maternal affection, but such is the case, and in a most remarkable manner; in fact, taking their young in their mouths, if alarmed.

Dr. Edward Palmer, a well-known traveler and collector, states that, when in Paraguay with the Water-Witch expedition, he saw seven young rattlesnakes (*Caudisona terrifica*) run into their parent's mouth. (*Plate XIX.*) After it was killed, they all ran out. These snakes, parent and brood, are preserved in the National Museum, Washington.

Palisot de Beauvois, an eminent French naturalist, thus details an observation made near the close of the last century: " When making my first excursion into the Cherokee country, I happened, while botanizing, to see a rattlesnake in my path. I approached as softly as possible; but just as I was about to strike, imagine my surprise to see it, after sounding its rattle, open a very large mouth, and receive into it five little serpents, each about the size of a goose-quill. Astonished at this singular spectacle, I retired some distance and hid behind a tree. After some minutes, the animal, believing itself out of danger, again opened its mouth and allowed the little ones to escape. I advanced; the little ones retreated to their stronghold, and the mother, carrying her precious treasure, disappeared among the underbrush, where I was not able to find her."

Among other rattlesnakes, the diamond attains a length of eight feet, and is strictly a Southern species, only found on the Atlantic south of the Carolinas. In the same locality is found the ground rattlesnake, about thirteen inches long. Others are the red, Mitchell's, horned, "side-winder," tiger, St. Lucas, banded, Arizona, spotted, scutulated, confluent, and the black-tail rattlesnakes, all of the genus *Crotalus*. In the *Caudisona* are found the Mexican ground, the Southern ground, the Sonora ground, and the black rattlesnake—a ferocious array. The bite of nearly all of these is extremely dangerous, though in the smaller it is not necessarily fatal. Almost all animals succumb to them, and man, if proper remedies are not at hand. It is a general belief that the hog is exempt; but this is at least not the rule, and where these reptiles are common, these animals are often killed by them.

Almost every locality has its seer with a remedy for snake bites; whiskey, however, is generally the most successful. This was well shown in New York in 1883, where a show man was bitten by a large rattlesnake, and recovered by being kept under the influence of large doses of whiskey. Being a temperance man, he asserted that he suffered more from the liquor than he did from the bite. This is one of the best authenticated cases of the utility of alcohol on record.

It is stated that M. Dr. Lacerda, of the French Academy of Sciences, has established the fact that *permanganate* of potash is the best antidote to the poison of snakes. The experiments were exhaustive, and deemed satisfactory. The scientist injected the active venom of a deadly snake, diluted with distilled water, into the cellular tissue and veins of a dog, and found that the antidote stopped the manifestations of venom injuries. The experiments were made in the presence of the Emperor of Brazil and a large company of scientists, and the fortunate experimenter has been decorated by the Emperor for his valuable discovery.

The juggling of the Hindoo fakirs with the deadly cobra, is familiar to all, and it is now known that it is done at times without resort to trickery; but it is not generally known that similar experiments have been tried with the rattlesnake in this country. Chateaubriand says: "One day, when we had stopped on a plain on the banks of the River Genedie, a rattlesnake entered our camp. We had a Canadian amongst us who played on the flute; wishing to amuse us he approached the animal with this new kind of weapon. At the approach of his enemy, the splendid reptile at once coiled itself up spirally, flattened its head,

PLATE XVI.

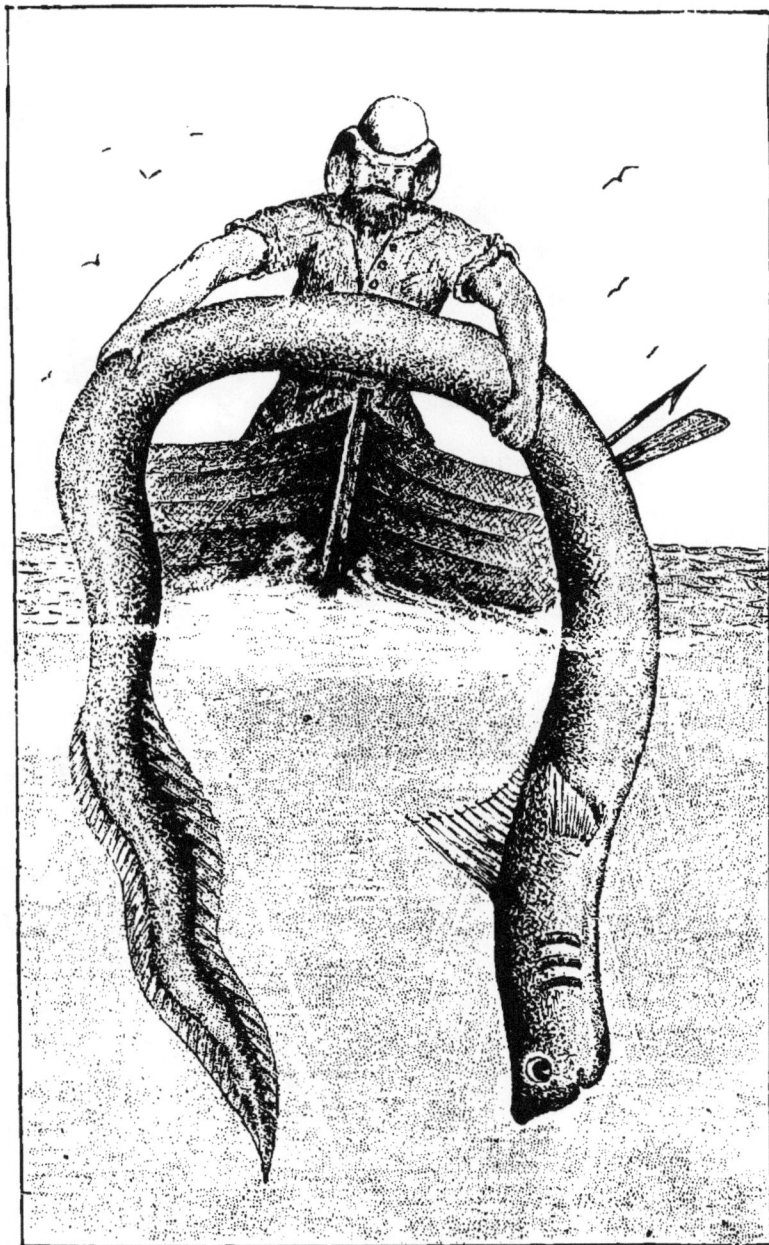

THE PEMAQUID SEA SERPENT.

puffed out its cheeks, and showed its envenomed fangs, while its forked tongue moved rapidly, and its eyes burned like red-hot coals ; its body became inflated with rage, rose and fell like a pair of bellows; its dilated skin bristled with scales ; and its tail which produced a sinister sound, oscillated with lightning rapidity. The Canadian now began to play upon his flute. The snake made a movement expressive of surprise, gradually drew its head backward, closed its inflamed mouth, and as the musical sounds struck it, the eyes lost their sharpness, the vibration of its tail relaxed, and the noise which it made became weaker and weaker, and finally died away altogether ; the coiled up line became less perpendicular, the coils of the charmed snake opened, and in their turn rested in wider concentric circles on the ground. The scales of the skin were also lowered, and immediately recovered their wonted brilliancy ; and, turning its head slowly toward the musician, the snake remained immovable in an attitude of pleased attention. At this moment the Canadian walked away a few steps, drawing low and monotonous tones from his flute ; the reptile lowered his neck, opened a way among the fine grass with its head, and crawled in the steps of the musician who thus fascinated him, stopping when he stopped, and following him when he began to move away. The snake was thus conducted from our camp, in the midst of a throng of spectators—as many red skins as Europeans—who could scarcely believe their eyes."

Of all the poisonous snakes of this country, the copperhead is the most dreaded. It is also known as the cotton-mouth, moccasin and red-eye in the South. It is common from the Catskill region to the Gulf States. It attains a length of two feet, is of a hazel hue, the head having a bright coppery lustre. The copperhead conceals itself in shady spots in meadows of high grass, feeding upon small animals, and rarely attacking large

9

ones unless stepped on ; in this way horses are sometimes killed by them. The mother copperhead has also been observed to take its young in its mouth when danger threatened them.

The water-moccasin (*Ancistrodon piscivorus*), from its pugnacious disposition, is perhaps equally to be feared. While the rattlesnake will slink away from danger, the moccasin will attack man or brute with savage ferocity, and thus is treated with respect by the negroes of the South. It is found from the neighborhood of the Pedee River to the Gulf States, and to some extent in the Mississippi Valley ; while another species—the *A. piscivorus pugnax*—is found in Texas, and called the Texas moccasin.

But the stronghold of the moccasin is in the vast swamp of Southern Florida, where the members of a recent expedition state that they exist in enormous numbers, having to be pushed aside as they advanced, and crawling into the boats and canvas at night. They are essentially water-snakes, chasing fishes and small reptiles in the streams of the rice plantations. They may be recognized by the dark-brown colors on the upper portion of the head, and the yellowish line extending from the snout to or over the nostril. They rarely attain a length of over twenty inches, and are short in proportion.

The moccasins show the same curious care for their young already mentioned. A few years ago, a gentleman, directing some hands at work on the lawn of Dr. Phares, of Woodville, Miss., heard a low, blowing noise, and on looking saw a large water-moccasin, and a large number of young hurrying to her head and disappearing so rapidly that he first thought they ran under her. He soon discovered that they went into her slightly-opened mouth, which was held close to the ground till they had all entered. She then attempted to escape, but was cut in two with a hoe; a number of young, eight or ten inches long were taken from her.

A gentleman in Georgetown, S. C., writes: "I had for several days noticed a very large moccasin coiled around the limb of a small tree near the pond. I concluded to capture it, and accordingly procured a large rabbit and placed it some way up from the pond, to toll her away from the water. She soon came down and disappeared under a large log; when next seen, she was near the bait, having traced it along the log on its opposite side. When she had nearly swallowed the bait, we made an advance; quickly disgorging it, she gave a shrill whistling noise, and five young snakes ran from under the log, and ran down the throat of the old one. We cut off her head and found the five young, which made efforts to get away."

One of the most beautiful snakes found in the United States, venomous, but not necessarily fatal in its bite, is the bead-snake (*Elaps*). It has permanently erect poison-fangs, but is extremely mild in its disposition. The coloring of the harlequin, as it is sometimes called, is exceedingly rich, the principal color being red, with seventeen broad, black rings, each bordered with yellow. It ranges from Virginia to Arkansas, four other species being known in Florida and Texas. The South American species are extremely dangerous. They are generally found underground, often being hoed up by the hands working in the fields.

One of the commonest of the non-poisonous snakes is the striped or common garter-snake, ten species of which are known in the United States. In early spring they are almost the first of the reptiles to roll out of their holes, where they have been hiding in balls or clusters. The author has seen them in New York before the snow was fairly away. Though easily aroused, and striking quickly, their bite is little more than a scratch. In the spring they are always hungry, and I have seen one chase a toad for fifty feet over a gravelly road, finally catching it. I captured the snake, and fed it with three adult toads within three hours. If the victim was seized by the head, his fate was soon determined; but when a hind leg was caught, the other, for a while seemed to offer an obstacle, and called to mind the fable that once went the rounds of the papers, to the effect that a toad seeing itself about to be swallowed, seized a stick and held it crosswise in its mouth, thus averting the danger.

The garter-snake is remarkably prolific, and in the spring their numbers about pools are sometimes astonishing. It has been suggested that they are both viviparous as well as ovoviviparous, from the fact that seventy-eight young have been taken from a single female, some being found free and others in sacs. With a brood of forty or fifty young, the striped snake would seem to have a difficult time in protecting its offspring by taking them in her mouth. They have this habit, however, as is shown by the following statement from a writer from Chesterfield, N. H. He says : "I saw a striped snake on the hillside, and noticed something moving about her head, and counted twenty little snakes from one and a half to two inches long. I made a move, and the old one opened her mouth, and they went in out of sight. I stepped back and waited, and in a few moments they began to come out. Then I made for the old snake, and killed her, and forced out several."

Another gentleman writes : "Some years ago I came across a garter-snake with some young ones near her. Soon as she perceived me she hissed, and the young ones jumped down her throat, and she glided beneath a stone heap. Another time I caught a snake of the same species, but, as I thought, of immense

size, which I took home and put in a cage. On going to look
at her some short time afterward, I discovered a great number of
young ones (about thirty, if I remember rightly), and whilst I
was still looking at the sudden increase, two more crept out of the
old one's mouth, and finally, after a little while, a third one did
likewise.

The black snake (*Bascanion constrictor*), that is a match for
the rattlesnake, often squeezing it to death, is a familiar form,
and widely distributed. A party of hunters recently observed
in Pennsylvania a black ball, two feet in diameter, rolling slowly
down a hill, and found on examination, that it was composed
of hundreds of these reptiles. In appearance they are very
attractive, from a decorative point of view, being of a steel-blue
uniform color, with a rich tessellated arrangement of scales. In
their natures, they are wild, untamable, powerful, and active
foes, often engaging in encounters with other snakes, especially
the rattlesnakes, quickly killing and forcing them to disgorge
their prey. In their movements they are so rapid that they are
often called the racer.

In the breeding-season they are bold, often going out of their
way to attack passers-by, and have been known to chase an
intruder for some distance. According to Holbrook, they will
even descend a tree to attack the one who is teasing them.

The black snake is the one that most frequently appears in
the guise of a charmer, but as I have previously suggested, this
power, so often imputed, is merely imaginary. The reptile
preys upon birds in their nests, penetrating thickets in search
of them; and often the cat-bird and the red-winged black-bird
are seen acting strangely, crying and fluttering before the reptile
in fear and rage, while thus *charmed*, frequently falling a victim
in their attempt to protect their young. At such times the cries

of distress of the old bird have called a number of birds of different genera, who assemble to join forces against the common enemy, finally forcing it to retreat.

Like other snakes mentioned, the black snake is said to protect its young by taking them in its mouth. The Rev. Chauncey Loomis, of Middletown, Conn., saw the Alleghany black snake open its mouth, and seven young pass down its throat, upon which the mother darted swiftly away.

In speaking of this interesting phase of snake life, Professor Goode says, referring to the testimonies on various genera: "The total number of testimonies in my possession is one hundred and twenty. Sixty-seven witnesses saw the young snakes enter the parent's mouth; twenty-two of these heard the young warned by a whistle, or hiss, or click, or sound of the rattles; five were considerate enough to wait and see them re-appear when danger seemed over, one seeing the act repeated on several days. These saw young snakes coming out of a large one's mouth, and not having seen them enter, were naturally much astonished. Five struck the parent, and saw the young rush from its mouth; eighteen saw the young shaken out by dogs, or running from the mouth of the dead parent. Thirty-six of those who saw the young enter the parent's mouth, found them living within its body. Only twenty of the sixty-seven allowed the poor, affectionate mother to escape. Thirty-three who did not see the young enter found them living within the parent's body."

The coach-whip snake, a long, slender form of the Southern and Gulf States, from its attenuation and rapid movements, has been vested with remarkable powers by the credulous. The Indians formerly asserted that it had the power of cutting its antagonist in twain by its whip-like motions; others thought that it formed itself into a wheel by holding its tail in its mouth and rolling away. Many of these curious fables are still believed in the South.

An interesting fact concerning the common milk snake has been shown by Professor C. F. Brackett, of Princeton, illustrating how, immediately after birth, the young make use of their instinctive faculty. He says : " A workman who was mowing in my father's hay field came upon a moist, moss-grown knoll, and his scythe cleft off a portion of the thick moss and sphagnum, and revealed several (at least a dozen, I should say) small soft bodies, which he declared to be snakes' eggs. I, at that time, having no knowledge of such matters, was incredulous, and proceeded to tear one of them open, when, to my surprise, there appeared a small, perfectly formed milk-adder, which immediately assumed a pugnacious attitude, and brandished its tongue as defiantly as an old snake would have done. Other eggs were torn open with like results. Soon the old snake appeared, and after endeavoring apparently to encourage the young family thus suddenly initiated into the world, it put its mouth down to the ground, and every one that had been liberated from the egg voluntarily and hastily disappeared within the abdomen of the old one. Last of all, I put the point of a pitchfork through the old snake, and with a pocket-knife opened the abdomen, and found the young ones still active."

The hog-nose snake, or blowing adder, common in the Eastern States, is, notwithstanding its disagreeable aspect, perfectly harmless.

Among the most attractive forms are the green snakes. The *Leptophis aestivus,* common in the South, is of a brilliant green color, and a perfect mimic of a vine, and would rarely be taken for a living creature when lying amid the branches of a tree. They have a habit of coiling in birds' nests, often surprising the egg hunter by bounding away. Allied species, further to the South, have been seen to leap twenty feet into the air, falling to the ground and escaping. They are perfectly harmless, and, like the common green snake of the North, can be handled with impunity; are easily tamed, feeding from the hand. A lady once appeared in public with three such pets,—one about her neck as a necklace, the others clasped about her wrists as bracelets.

The water-snake (*Nerodia sipedon*) is an extremely common form in the Eastern United States; and an allied form, known as the red-bellied water-snake, is common in Michigan, while several other species are well known in various localities. They are inoffensive creatures, and prey upon small animals. In Ohio they are quite common, and a farmer of Mechanicsburg, Ohio, says that, seeing a large one on Deer Creek, he procured a pole for the purpose of killing her. One stroke slightly wounded her, and she immediately made for the water. After she had swam about her length, she wheeled, placing her under jaw just out of the edge of the water, and opened her mouth to the fullest extent. Some dozen young snakes, three to four inches long, then seemed to run, or rather swim, down her throat, after which she clumsily turned in search of a hiding-place. He opened her, and found about twenty living young snakes, two or three, seven or eight inches long. This shows them to have the same peculiar habit noticed in so many others.

The water-snakes are generally found about water-courses, often hanging in the branches of trees over streams, into which they rapidly drop. Dr. Bell, the English naturalist, tamed a European species of this genus. His pet would distinguish him among a crowd, and crawl to him, passing into his sleeve, where it curled up for a nap. Every morning it came to the doctor's

PLATE XVII.

ELASMOSAURUS PLATYURUS—FIFTY FEET IN LENGTH.

table for its share of milk. For strangers it had an aversion, flying and hissing at them when touched.

In tracing back the history of the snakes, we find that the earliest occurs in the Eocene Tertiary period, and their fossil remains are found in various parts of the country. At this early time, several species of a great water-snake, at least eighteen feet in length, lived on the New Jersey coast. Remains of others, allied to the existing boa-constrictor, have been found in the Eocene lake deposits of the West; so, comparatively speaking, the snakes are of modern origin.

CHAPTER XII.

ANIMAL MIMICS.

A NUMBER of years ago reports came to the headquarters of an English regiment stationed in India, that a portion of the country lying in its jurisdiction was being systematically pillaged by a band of native robbers. As soon as practicable a force was sent against them, and for many days the trail was followed without success. One evening when the troops had gone into camp after a long day's march, one of the men led his horse toward some curious blasted and upturned roots that stood upon the plain near by, thinking to there tie the animal that had shown a disposition to run away. The private had approached within a few feet of the roots, when to his amazement, they moved, actually turning into men, who, with derisive cries, darted away, disappearing in a rocky defile where they were safe from pursuit. The wily natives had, by twisting and contorting their bodies, so mimicked the roots of trees that they would have escaped observation but for the accidental approach of the soldier.

In a very similar way many of the lower animals are enabled to avoid their enemies or better approach their prey. Who is not familiar with the protection afforded our game-birds by the adaptation of their colors to those of their surroundings? Quail, woodcock, and grouse have almost the exact colors of the grasses they most affect. In the North we find the ptarmigan in winter assuming a white garb, so that against the snow it cannot be distinguished; and as the summer months come on, its plumage changes, and is adapted to the gray lichens and mosses of the warmer times. The polar bear, and many of the hares, gulls, and carnivorous animals of that region, have a white coloring. The young of some seals are pure white, a color that serves them well; and deer, elk, and the larger game offer no strong contrast as they stand amid the solitudes of their choice.

In the great arid tracts of the West, where the sun has given everything a dry, withered tint, we find similarly-colored animals. The horned-toad so resembles the ground that it is hardly noticeable. The tawny skin of the lion, the stripes of the leopard, the gray color of our puma, that renders it inconspicuous when crouching on a rock, are all familiar examples of this natural endowment.

These, however, are only generalized cases, and might be passed as accidental occurrences were there not instances so wonderful that design is evident at the very outset. Among the fishes they are extremely numerous. This is especially noticeable in the family of which the goose-fish, or angler, is a member. In drifting along we see upon the bottom a black moss-covered rock; the kelps and algae that grow upon it wave to and fro in the current, and not until the apparent rock is touched do we discover that it is a fish. The huge mouth is fringed with barbels of flesh that are perfect imitations of the local weed in shape and color; the pectoral fins look like sprays of it; while the curious dorsals are covered with dancing filaments that are seriously believed by many a luckless fish to be marine plants. The most remarkable mimic of this family was discovered by the "Challenger." The entire fish was apparently covered with fringes and streamers of flesh that gave it the exact appearance of a bunch of sea-weed.

In Eastern seas we find the grass-fish (*Nemichthys*), which is invariably seen upright among the grass it resembles. The Antennarius and Chironectes resemble the gulf-weed in which they live not only in form, but in color; yellow, green, and white tints are scattered over their bodies in seeming confusion, so that two alike have never been found. Even the white polyzoons that grow upon the sargassum appear to have been imitated, and thus, lying partly submerged upon the gulf-weed, exposed to the watchful eyes of predatory gulls, these fishes find absolute safety in the resemblance.

The sea-horse (*Hippocampus*) and pipe-fish are wonderfully protected. In a recently discovered species of the first, the fins seem to have been replaced by long streamers, in some cases as long as the fish itself, that wind about, moving to and fro in the water, so that when clinging to the weed by its prehensile tail, and surrounded by its waving fins or filaments, its most inveterate enemy would pass it by as a mat of sea-weed. The pipe-fishes may be found in the eel-grass near the mouths of the Delaware, Hudson and other rivers; and so closely do they resemble bits of weed that the keenest observation is required to find them, and often the movement of the fish as the grass is lifted from the water is the only means of detecting the little mimics.

Among the fishes that rely upon color alone as a protection, might be cited the little Nomeus, as showing what may be called the perfection of mimicry. I first observed these fishes in the Gulf Stream, in the vicinity of Yucatan, in 1860. The water was covered with myriads of the fairy physalia, or Portuguese man-o'-war, which, with their oval hulls and pink and silver sails, were cruising about at the mercy of the warm trade wind. Wishing to obtain specimens, we rowed among them. Upon lifting one from the water, which could be done safely by the sail, several dark fishes were observed darting wildly about.

Thinking they might be attendants upon the physalia, we returned the latter to the water, and no sooner was this done than four or five little fishes of the most intense blue darted to it from under the boat, and took their places under the long, blue, deadly tentacles. They were of the same color, so that within a short distance of the floating bubble the fishes could hardly be distinguished. But even more remarkable than this resemblance was their immunity from the death-dealing lasso-cells of their protector. I have seen a sardine touch one of the tentacles and succumb as if by an electric shock; yet here in the midst of the virulent blue stings the little blue fishes found a home. Evidently the powers of the physalia are well known to the inhabitants of the sea, and they give it a wide berth; but the blue fishes, by being mimics of the tentacular parts of their host, find protection. Jelly-fishes were often seen side by side with the physalia, also having their attendants, who imitated their host in being transparent and tipped with reddish and pink tints; yet never did we find the blue friend of the physalia under the jellies, or *vice versa.*

The perch, dolphin, stickleback, corbitis, etc., have long been noted for their power of changing color, some assuming a new tint instantaneously and others more gradually. If we arrange small enclosures in shallow water, having the bottoms of each different, one pure white sand, one covered with black kelp, and another a medium shade, we shall find, if we place flounders of the same general tint in each, that they will very shortly assume a color assimilated to that of the bottom. Now take out the flounder that has been on the white bottom and place it on the black kelp, and it soon appears to observe the difference, gradually adopting the color. If a blind flounder of normal color is placed on the dark or white bottom, we find that no change is

effected. The conclusion would naturally be that the fish had
observed the change that brought it out in such bold and dangerous relief, and had exerted certain powers of reason to produce a like change in itself. This, according to Semper, does not
seem to be the case. We find that the skin consists of two distinct
portions, the epidermis and the cutis, the former being entirely
composed of cells, the innermost layer containing cylindrical ones.
The cutis is composed chiefly of fibre, and encloses nerves and large
cavities for glands and cell-elements. These contain pigment,
and to its distribution and the power of the highly-ramified cells
of shrinking and expanding under various circumstances are due
the remarkable variations of color that we are all more or less
familiar with. The pigment in the cells varies in different fishes
or animals, and in different parts of the body, being yellow,
black, red, green, or brown, as the case may be. In experiments
with the goby it has been found that the pigment-cells, or chromataphores, that are yellow or greenish-yellow when distended,
become orange-colored when contracted, while the orange or red
cells, when exposed to an irritation that produces shrinkage,
become black, or even brown. The pigment-cells are arranged
in layers in the cutis; close to the epidermis are the light-
colored yellow cells, beneath them the red or brown, below, the
black. If they all relax, then the prevailing color will be black
or brown, and the patches of light-colored cells will be dulled.
If they contract while the light patches are yet expanded, the
latter will be rendered more conspicuous, and the fish or frog
will appear mottled. As we have seen, the eye is the medium
by which the irritant is conveyed, and the latter is undoubtedly
reflected light, producing certain effects upon the optic nerve
that are transmitted to the sympathetic nerves extending parallel
with the vertebrae, thus reaching the chromataphores. That

this is the correct explanation can be shown by severing one of the nerves, and by careful manipulations one side of the fish may be made to appear striped, while the other will retain its normal tint; so it will be seen that the immediate accomplishment of a protective resemblance is an intuitive change, and does not involve any mental action on the part of the animal, which is perhaps unconscious that such a change has taken place.

An instance of involuntary adaptation occurs among the shrimps, similar to that of the flat fish cited. The chameleon shrimp is found generally among grasses of vivid green, with which it agrees in color. If we remove it to a jar of water, it grows pale, the green tint finally disappearing, until we have a transparent shrimp, almost invisible. On brown sand it also assumes that tint; but when its eyes are destroyed it will be found to assume any color, as in the case of the blind flounder.

Among the crabs are a number of striking mimics. Our familiar rock crab, if taken and cleansed (and I have often tried the experiment), will immediately, when placed in the water, commence to plant upon its back bits of weed that it cuts from the bunches all about. The severed branch is first placed against the mouth and some glutinous secretion affixed, then placed upon the back, where it lives and grows. In less than an hour one of these crabs will cover itself so that only sharp eyes can distinguish it from a moss-covered stone. Others, as Parthénope, resemble rocks in their structure, so that many a skate or codfish would pass them by not suspecting that its legitimate prey was concealed beneath the rough exterior. I have seen a large land-crab in the South ascend the prickly pear, evidently to feed upon the fruit, and so imitate it in color and shape that at a few feet it was absolutely unnoticeable, and when alarmed it would draw back into the crevice and assume almost the exact position of the fruit. These crabs were found in great numbers on the fallen mangrove-boughs that covered the beach ; but on the pure white shore, where their rich purple bodies would soon have caught the eye of the watchful laughing gull, we never saw them. There were crabs here, however, in great numbers, but every one was perfectly white, the same shade as the sands bleached for years in the tropical sun.

How interesting is the mimicry of the little isopods, so common on rocky shores ! Every stone overturned reveals myriads of them, no two alike, decked in varied garbs, blocked and dashed with grotesque colorings in imitation of the barnacles and weed that hang upon the rocks about them. One, the *jæra copiosa*, abounds in greater numbers than the larger forms, and, with its shades of gray, slate, green, brown, and white, successfully simulates its surroundings. So with the Idotea, found clinging to the eel-grass, or ulva, of which it seems a veritable part.

Among the hydroids and delicate forms of algae, we shall discover the curious capsella clinging to the surface with its stick-like legs, safe in its protective garb and shape. The most remarkable crab mimics are found in the Sargasso Sea ; and in many I have caught, from the delicate *Nautilographus* to the larger and more predatory forms, all had fantastic markings exactly resembling the leaves of the sargassum. The backs and claws of the larger crabs looked, when first taken from their resting-place, as if another artist than nature had been at work. The yellow and delicate brown and green tints were perfectly reproduced, even the membranipora that grew upon the weed seeming to be imitated. How necessary is this protection, is seen in the numbers of gulls that roam over the vast areas of sargassum and would be attracted by the slightest contrasting feature. This surveillance on the part of the birds has undoubt-

PLATE XVIII.

AN EXTINCT MARINE REPTILE, CAMARASAURUS—SEVENTY-FIVE FEET LONG.

edly resulted in the perfection of mimicry. The crabs that offer strong contrast have always been observed and devoured, only those with protective tendencies being preserved, whose progeny would naturally in each successive generation become more and more in keeping with their surroundings, assuming that the unprotected exceptions were being continually weeded out, which is probably the case.

Not only the crabs, but all the inhabitants of the gulf-weed are equally protected. On the shrimps the imitation of the bryozoon *Membranipora* is wonderful. The planarian worms cannot be distinguished from the weed itself. The mollusk *Scylloca* not only has the prevailing color, but its members and body seem to have been modified even in form to suit the exigencies of the case. Among the weed is found the fairy-like Ianthina, a beautiful blue shell, and clinging to its raft we find the little crab *Nautilograptus,* here a perfect blue, while but a moment before we have seen others upon the weed, their backs white and yellow.

All the pelagic animals are either colorless or specially provided with a protective tint. The Ianthina and Vellela are both blue,— the rich tint of the Gulf Stream; while the floating raft of the former we should pass by as a fleck of sea-foam. The dark tint of many of the *Salpae* causes them to resemble bits of sea-weed, though at night they blaze with unwonted splendor—the most beautiful of the luminous forms. Perhaps below the floating weed, with its pelagic inhabitants, we shall find the sea-cucumber—the very prince of mimics. In some the organs resemble the leaves of the pineapple; in others the tentacles exactly imitate some delicate branching algae; while again, toadstools seem growing from the sand. The common sea-cucumber (*Pentacta*) of our shore can easily be examined. Its black, leathery, bag-like

10

body offers a strong contrast to the sand—a fact seemingly known to it, as when dropped into the aquarium it almost immediately disappears. It is not long, however, before a shapely tendril breaks through the sand; soon another and then another appears, until a perfect shrub is seen waving in the current, and hardly noticeable among the branches of real weed around it: these are the breathing-organs of the Pentacta.

INSECT MIMICRY.

Perhaps, however, the most striking instances of mimicry occur among the insects. Some resemble leaves, sticks, and other inanimate objects; others find safety in imitating insects that are known to birds as being poisonous, and are hence avoided by them. The walking-sticks are curious examples of the first class, resembling the twig upon which they rest, not only in color, but in their very structure and joints. Some found in the tropics are eight inches in length. The decayed particles of a dead limb, the small rugosities of the bark, the moss or fungus that may grow upon it, are perfectly reproduced in these curious insects, which, hanging from the limbs with legs all awry, would never be considered living creatures by the casual observer. (*Plate XX.*) The leaf-insect (*Phyllium*) (*Plate XXI.*) is equally wonderful, in shape resembling a moderate-sized leaf; the veining and tint are near copies of the leaf upon which they live; and travelers have touched them without finding out the deception. In Java they are kept alive, and strangers are often asked to point them out, and though numbers are before them, they find it difficult to detect the mimics from the true leaves. To make the resemblance more perfect, many have mould spots, or what appear to be such, upon their wings. This is still oftener the case with the locusts, many of which bear a strong resemblance to leaves. Mr.

Belt, the naturalist, observed the use of this resemblance in a case where one of the mimic locusts was overtaken by an army of ants. The locust might have flown away, but it would then have been devoured by the birds that were preying upon the ants; so it merely remained quiescent, imitating an old worm-eaten dried leaf, and was passed over unnoticed by the swarm of ants, after which the cunning mimic moved off. In Java, the home of the beautiful leaf-butterfly, so noted for its curious forms, a pink mantis has been discovered that resembles a pink orchis so closely that it is difficult to distinguish them. Here the insect is protected from the birds, and, being carnivorous, it is enabled to capture other insects that mistake it for a flower, and is thus a self-baited trap.

Many insects that are harmless and unprotected are found to mimic in shape and color, other insects that are notably aggressive. Thus, the stinging ants in Central America are imitated by spiders and certain beetles; a virulent stinging wasp, which few birds would venture to touch, is found to have a double that is in reality a harmless bug. Many flies resemble bees, both in appearance and in the sounds they produce. The butterflies, as *Danais* and *Archippus*, that are distasteful to the birds, perhaps on account of their poisonous nature, are mimicked by butterflies that the birds are eager for. Mr. Wallace observed a remarkable case of this in Malaisia. He noticed a butterfly which the birds always recognized and avoided, mimicked by an *eatable* butterfly. As soon as the bird started in chase, the wily insect assumed the clumsy and laborious movement of the wings that served as a danger-signal to the other species, and the bird at once gave up the pursuit.

One of the most remarkable facts in this phase in the struggle for existence is that not only do the insects adopt the color about them, but the chrysalis of an African butterfly (*Papilio niveus*) has the same power. This was first noticed by Mr. T. W. Wood, who found that the colors of the chrysalis of the small cabbage-butterfly changed according to the tint of the interior of the box in which it was confined. Those in white boxes were white; those in dark, black; some against a brick wall were reddish. The same is true of the cocoon of the Emperor moth.

What wonderful appreciation of the benefits of a protective resemblance is seen in some of the trap-door spiders! If left as completed, the oval door of silk would attract enemies; but to avoid this the cunning builder brings earth, places it upon the trap, and here plants moss and bits of weed, which it brings from far and near, that soon take root and grow, effectually concealing the door and trap. Other species as the *Nemesia*, fasten entire leaves upon the top of their nests, and thus disguise them.

Among the larval and pupa forms of insects, the examples are both numerous and wonderful. Some resemble bits of stone; others small shoots of trees just starting, the buds of flowers, and even the flowers themselves. Others again mimic bits of moss and various objects of different colors. The caterpillar of *Coelodasys unicornis*, common on hazel leaves, is an interesting mimic. It eats into the edge of the leaf, extending the body along the eaten portion, following the curve and clinging to it by its feet, and in this position is very difficult to distinguish. A species of *Notodonta*, that is often found on the willow, is a remarkable mimic, and resembles the rugosities of the tree.

A MIMICKING CATERPILLAR.

Mr. S. E. Peal reports from Asam a very interesting case of mimicry in a caterpillar. The animal, when at rest and undisturbed, resembles the upper outline in *Plate XXII.*, and is a large insect of a general brownish gray color with green markings. When attacked or alarmed, it immediately assumes the ferocious appearance indicated in the lower figures of the cut, and becomes in a second a mimic of an insectivorous animal, a shrew. So striking is the resemblance that Mr. Peal was completely deceived. Only the fact that it did not run at his approach caused him to examine it closer, when he found instead of a shrew a harmless caterpillar that probably thus mimicked the very animal that perhaps preyed upon it.

The rich green iguanas are almost invisible among the leaves of their native haunts; the bull-frog among the sedges by the river-side is green also; while the toad that affects dusty roads is mottled or gray.

A wonderful mimic is the chameleon, though our *Anolis* is equally interesting. In this case, according to an American naturalist, the color of the integument can seemingly be altered at the will of the animal. In it the two layers of movable pigment-cells are deeply seated in the skin, but can be brought to the surface. The layers are blue and yellow, and by pressing certain muscles they can be forced up either separately or together; when the muscles are not brought into use, the general color of the animal is grayish; but when occasion requires, it applies the lever, if we may so call it, and by an action of the will greater or less pressure is brought to bear, producing tints varying from blue or green to yellow, brown, or mottled.

Who can doubt that the bits of moss placed by the humming birds upon their nests, rendering them so difficult to find, are intended to have this result? The nests, when completed, are perfect imitations of the moss-tipped knobs of the apple tree, and are often passed by unseen by those in search of them.

The curious Australian bird, Megapodius, when pursued or in danger, has been seen to alight upon a limb and straighten out its neck, so closely resembling a branch of the tree that the pursuer was completely discomfited.

Even among plants we find strange cases of similar means of protection, that of the stone mesembryanthemum of the Cape of Good Hope being one of the most striking. It so resembles the stones among which it grows that the novice is always deceived and passes specimens by when making most careful search. The little plant is extremely juicy, and especially coveted by grazing animals of all kinds, and would undoubtedly soon be exterminated were it not for the singular protection that causes it to be mistaken for stones. Other cases are known of curious plants growing in the stony karroo, with the tuberous roots above the soil, so that when not in leaf their resemblance to the surrounding rock is perfect. Some of the fungi are mimics of other forms that are known to be unsuitable for food, and thus find protection. The labiate plant *Ajuga ophrydis* of Southern Africa mimics an orchid, and thus may attract insects to fertilize it.

Such are merely a few out of myriads of examples that illustrate one of the many devices employed by nature in preserving her offspring.

CHAPTER XIII.

Not the least interesting feature connected with the extinction of animals, is that relating to the manner of their disappearance, and the conditions that caused these myriads of forms to be swept away, leaving only their hardened remains to tell the story. Without referring particularly to the older geological ages, let us glance at some of the more recent forms that have disappeared within the time of man.

At the discovery of St. Helena, Deadwood and Longwood, so well known from Napoleonic associations, were covered with forest-trees. In 1502 goats were introduced on the island, and eighty-six years later had increased to great numbers, and were eating the young trees, the old ones rapidly falling from age. In 1731 the inhabitants were aroused at the prospect of the total destruction of their forests, and all stray animals were killed,—too late, however, to save Greatwood, as the wooded country was called, and in a short time nearly every tree had disappeared, the entire tract now being pasture-land. With the destruction of the forest came that of a multitude of insects dependent upon such conditions, and many land-snails, in all eight species, were entirely exterminated in a period of about two-hundred and twenty years, and, as the snails were peculiar to the island, the extinction was complete. A similar fate befell many of the animals of the Virgin Islands, where the crews of vessels set

151

fire to the bush, thus destroying the vegetation upon which the animals were dependent.

Examples of extermination in localities are common on our own shores. The oyster beds once frequent upon the coast of Maine are now represented only by the shells piled in heaps along the shore.

EXTINCT ELEPHANTS.

The disappearance of larger animals seems perhaps more inexplicable; yet that early man is responsible in a greater or less degree for the extinction of the mammoth there can be but little doubt. From the earliest times the remains of these elephantine giants have been found in the North. One of the most important discoveries was made at the mouth of the Lena River, Siberia, at the beginning of the present century. A fisherman first observed the monster projecting from the frozen ground or tundra. Each successive year it became more exposed, and finally fell upon the sand as fresh as though it had died a month before, although in the estimation of some naturalists the last mammoth's death occurred at least a million of years ago. The body was somewhat damaged by bears and foxes, but the skeleton and part of the hair were preserved and are now in the museum in St. Petersburg.

Another mammoth was unearthed by a land-slide in 1839 on the shore of a lake near the mouth of the Yenisei. It was extremely perfect, and, according to the natives, a black tongue (the trunk ?) as long as a month-old reindeer calf was hanging out of its mouth. In 1842 it was secured by a merchant, but had been badly torn.

The remains of mammoths are common near the shores of the Polar Sea, and especially on the New Siberia Islands, which

PLATE XIX.

YOUNG RATTLESNAKES ENTERING THEIR MOTHER'S MOUTH.

appear to be a vast burying-ground for these monsters. The greater number probably died a natural death; or, perhaps wandered out upon the floes (*Plate XXIII.*), becoming entombed in the ice, so falling victims to the polar bear and other animals of prey. But within a few years similar remains have been found in France in company with flint arrow-heads and roughly-worked pieces of ivory bearing rude drawings of the animal, facts that point conclusively to man as their contemporary and undoubted enemy.

Many other animals have been found mingled with the remains of man, who is still in the ascendant, while they have passed away. Near Aray, in the department of Aube, France, the jaw of a human being has been found in a mixture of bones of the great cave-bear, hyena, and rhinoceros; while in Kent's Cavern, near Torquay, England, and in other places, flints and rude pottery have been found mingled with the bones of ancient and extinct mammals. One of the most remarkable finds was made in a cave near Aurignac, in the Pyrenees. Here human skeletons were discovered with flint and bone implements, together with fragments of the cave-bear, cave-dog, wild boar, bison, Irish elk, reindeer, and many others that had been carried in by their rude captors and used as food, the bones of many showing where they had been split to obtain the marrow. In various countries such finds have been made, showing that the extinction of large and conspicuous animals has been hastened by human intervention.

In 1742 Behring's Island was inhabited by herds of sea-cows, or manatees, that attained the enormous growth of twenty-five or thirty-five feet, and weighed five or six tons. (*Plate XXIV.*) They were of a dark-brown color, streaked with grayish or light stripes; the skin was thick and leathery, protected by a dense growth of hair that formed an exterior protective skin resembling the rugged bark of a weather-worn tree. Instead of teeth they had two masticating plates, one in the gum and the other in the under jaw. Vast herds of them were discovered by Steller, who, with a ship-wrecked crew, visited the island in 1742. They were found feeding upon the fields of sea-weed, that skirted the shore, and when attacked, showed a remarkable attachment for one another. Warfare was waged against them by all comers, with such effect that twenty-seven years later they were nearly all extinct, and now not one exists. Nordenskiöld thus refers to this great animal: " I succeeded in actually bringing together a very large and fine collection of skeleton fragments. When I first made the acquaintance of Europeans on the island, they told me that there was little probability of finding anything of value in this respect,—for the company had offered one hundred and fifty roubles for a skeleton without success. But before I had been many hours on land I came to know that large or small collections of bones were to be found here and there in the huts of the natives. These I purchased, intentionally paying for them such a price that the seller was more than satisfied and his neighbors were a little envious. A great part of the male population now began to search for bones very eagerly, and in this way I collected such a quantity that twenty-one casks, large boxes, or barrels were filled with Rhytina bones, among which were three very fine complete skulls, and others more or less damaged, several considerable collections of bones from the same skeleton, etc. The Rhytina bones do not lie at the level of the sea, but upon a sand bank thickly overgrown with luxuriant grass, at a height of two or three metres above it. They are commonly covered with a layer of earth and gravel from thirty to fifty centimetres in thickness. In order to find them, as it would be too troublesome to dig the whole of the grassy bank, one must examine the ground with a pointed iron rod, or bayonet, or some such tool. One soon learns to distinguish by the resistance and nature of the sound whether the rod stuck into the ground has come into contact with a stone, a piece of wood, or a fragment of bone. The ribs are used by the natives, on account of their hard ivory-like structure, for shoeing the runners of the sledges or for carvings. They have, accordingly, been already used upon a large scale, and are more uncommon than any other bones. The finger-bone, which perhaps originally was cartilaginous, appears in most cases to be quite destroyed, as well as the outermost vertebrae of the tail. I could not obtain any such bones, though I specially urged the natives to get me the smaller bones, too, and promised to pay a high price for them."

The disappearance of such large animals and in such vast numbers in so short a time seems incredible; yet, without governmental intervention, the sea-bears of the far North will soon be extinct, over three and a half million skins having been imported from the Pribylov Islands alone in eighty-four years, while elsewhere for many years the slaughter has been carried on without restriction.

The extinction of the great auk is undoubtedly due in a great measure to man. In 1834 Nuttall wrote: "As a diver he is unrivalled, having almost the velocity of birds in the air. They breed in the Faroe Islands, and in Iceland, Greenland, and Newfoundland, nesting among the cliffs, and laying but one egg each. They are so unprolific that if this egg be destroyed no other is laid during that season. The auk," he continues, "is known sometimes to breed in the Isle of St. Kilda and in Papa Westra; according to Mr. Bullock, for several years past no more than a single pair had made their appearance." Now there is not a single living species, and specimens are so rare that the Museum of Natural History, Central Park, paid six hundred and fifty dollars for one and the cast of an egg, the only others known in this country being in the cabinets of Vassar College, the Philadelphia Academy of Sciences, the Smithsonian, and Cambridge University. Not many years ago, comparatively speaking, they were common as far south as Nahant on the New England coast, and the great shell-heaps of Maine have furnished many of their bones. They also occur in the Danish kitchen-middens, and deposits containing them have recently been found at Caithness.

The great auk was commonly known as the gare fowl, and the last living specimens were killed in 1844, at a group of islands called Funglasker, off the southwest coast of Iceland. In 1833 a fine one was taken by some fishermen at the entrance of Waterford harbor, another in 1821 near St. Kilda, while a pair killed at Papa Westra, in 1812, now adorn the ornithological cabinet of the British Museum. In the eighteenth century these birds were common in the Faroes, and in the Iceland seas there are three localities named after them, so numerous were they, and now the name and tradition alone tell their story. It would seem that they were gradually driven from one nesting-place to another, and from the earliest settlement of the country. In 1813 the sailors of a Faroese craft, after successfully driving them from the open shores, followed them to a rookery formerly considered inaccessible, and destroyed great numbers of them. Seven years later Faber the naturalist attempted the same feat, but failed. In 1830, as if nature herself was in league with man against the birds, the inaccessible skerry, by a submarine eruption, was engulfed by the sea, the survivors establishing a colony on a rock called Eldey, near the mainland. During the following fourteen years sixty birds and eggs were taken, and finally, in 1844, the last pair were destroyed. In Newfoundland the birds were known as penguins, and were followed with equal pertinacity. In 1536 the French and English vessels drove them ashore or into their boats in droves, or "as many as shall lade her," salting them down as provision; and it would seem that the French rarely provisioned their vessels with fresh meat, depending entirely upon the auks. On Funk Island can be seen to-day rude enclosures made of stone, in which the luckless birds were imprisoned previous to slaughter—monuments of a lost race.

Found in the same localities, and meeting a common fate, was the Labrador duck—a large, handsome bird, which, as late as thirty-five years ago, was quite common in summer months about the mouth of the St. Lawrence and the eastern shores of Labrador, finding its way in winter along the coast of Nova Scotia, New Brunswick, and New England. Its size and appearance made it a strong attraction to the sportsman; its eyries on islands safe from foxes were sacked, and finally, driven to the shores of the mainland, where the eggs became the prey of predatory animals, it gradually succumbed, and the last one was killed by Colonel Wedderburn in Halifax harbor in 1852.

The disappearance of the gigantic pigeon-like bird *Didus ineptus,* commonly called the dodo, is no less remarkable. When the Portuguese under Mascarenhas discovered Mauritius in the early part of the sixteenth century, the bird was extremely common there; but not until Van Neck's voyage in 1598 was a definite account given of them. The Dutch call them *walgh rogels,* meaning nauseous birds, and this quaint description is given by Brontius: "The dronte, or dodors, is, for bigness, of mean size, between an ostrich and a turkey, from which it partly differs in shape, and partly agrees with them, especially with the

African ostriches, if you consider the rumps, quills and feathers, so that it was like a pygmy among them, if you regard shortness of legs. It has a great ill-favored head, with a kind of membrane resembling a hood; great black eyes, a bending, prominent, fat neck; an extraordinary long, strong, blue-white bill, only the ends of each mandible are a different color—that of the upper black, that of the nether yellowish—both sharp-pointed and crooked; its gape huge, wide, as being naturally voracious. Its body is fat and sound, covered with soft gray feathers, after the manner of an ostrich's, on each side. Instead of hard wing-feathers or quills, it is furnished with small soft-feathered wings of a yellowish-ash color, and behind, the rump, instead of a tail, is adorned with five small curled feathers of the same color. Four toes on each foot, solid, long, as it were, really armed with strong black claws." He adds, in conclusion, "It was much more pleasing to the eye than the stomach." The literature upon the subject is extremely voluminous, but unsatisfactory in detail. The first English observer of the bird was Emanuel Altham, who, in a letter to his brother, says: "You shall receive . . . a strange fowle, which I had at the Iland Muritius, called by ye Portuigalls a Do Do, which for the rareness thereof I hope will be welcome to you, if it live." Whether the bird was received is not known; but Herbert, who sailed in the same fleet, refers to the bird in the following: "The dodo comes first to a description here, and in *Dygarrois* (and nowhere else, that ever I could see or heare of) is generated the Dodo (a Portugnize name it is, and has reference to her simpleness), a bird which for shape and rareness might be call'd a Phœnix (wer't in Arabia.)" He also gives a quaint figure of the bird.

Cauche, who made the voyage to Mauritius in 1651, says that the dodo had a cry like a young duck ("*il a un cry comme l'oison* "), and that it laid a single white egg, "*gros comme un pain d'un sol,*" on a mass of grass in the forests.

The dodo was undoubtedly sent many times to Europe alive. Sailors, we are informed, killed them to obtain the stones in their crops, upon which to sharpen their knives, and finally they totally disappeared—a few pictures, a foot in the British Museum, a head and foot at Oxford, a perfect skull at Copenhagen, and a fragmentary piece at Prague, being all that is left to attest the reality of the existence of the king of the pigeons, the last of which were probably destroyed in the beginning of the eighteenth century.

To the south of Mauritius lies a small island which has successively born the names of Mascarenhas, Bourbon, and Réunion, and was formerly the home of a large bird known as the solitaire. Du Bois, in 1674, gave a meagre description of it, and Boutekoe and Witthoos have left rude sketches of it.

On the island of Rodriguez lived a didine bird, the *Pezophaps solitarius* of Leguat, a Huguenot exile, who lived on the island for some time (1691–1693). He left an account of the birds, and several cuts, whose authenticity has been amply shown by the discovery later of nearly perfect skeletons of the birds, which may now be seen in the museum at the University of Cambridge, England. These, however, by no means comprise all the last forms of the Mascarene group. In the library of the German emperor is a picture of a long-billed, flightless ralline bird, known as the *Aphanapteryx*, which at one time lived upon one of the islands, and recently its bones have been found in the peat of the Mare aux Songes. Here were also two species of parrot, a dove, and a large coot, the remains of which are to be found in a few scientific collections. In Réunion a starling, *Fregilupus*, with a beautiful crest, existed until about forty-five years ago, when it became extinct. In Rodriguez nearly the entire original *avifauna* has disappeared, including a large heron, a small and extremely peculiar owl, a parrot, and a dove.

A parrot (*Nestor productus*) has recently become extinct at Phillip Island, near New Zealand, the last living specimen having been seen by Mr. Gould in 1851, and hardly a dozen specimens of skins are known in all the museums of the world.

New Zealand is pre-eminently the land of gigantic extinct birds, which were undoubtedly destroyed by the natives, their roasted remains, egg-shells, etc., being found, together with those of human beings, showing that they formed part of a cannibalistic feast indulged in by the ancestors of the Maori. The most important discoveries of these giant-birds have been made in caves by Julius Haas, F. R. S., and the results of his labors are seen in the Smithsonian and the New York Museum of Natural History, the latter possessing the finest collection known.

That the gigantic moa was destroyed by man is proved also by the Maori traditions which tell of the great birds, and the songs that extol the magnificence of their plumage and the skill of their hunters, who ate the birds, reserving their wonderful plumes as decoration. These, and the unequivocal marks of the celt or jade axe on the recent leg-bones, are incontrovertible proofs that the giant moa, like the dodo, gave way before the advance of higher and human animals.

In 1847 Mr. Walter Mantell found the remains of a bird mixed with those of the moa at Waingongora, New Zealand. It was named the *Notornis*, and described by Professor Owen as a gigantic extinct rail, and in 1850 the soundness of his physiological inferences and deductions were shown by the discovery of a living *Notornis*,—a magnificent creature. Since then one or two other specimens have been killed, and now doubtless the

PLATE XX.

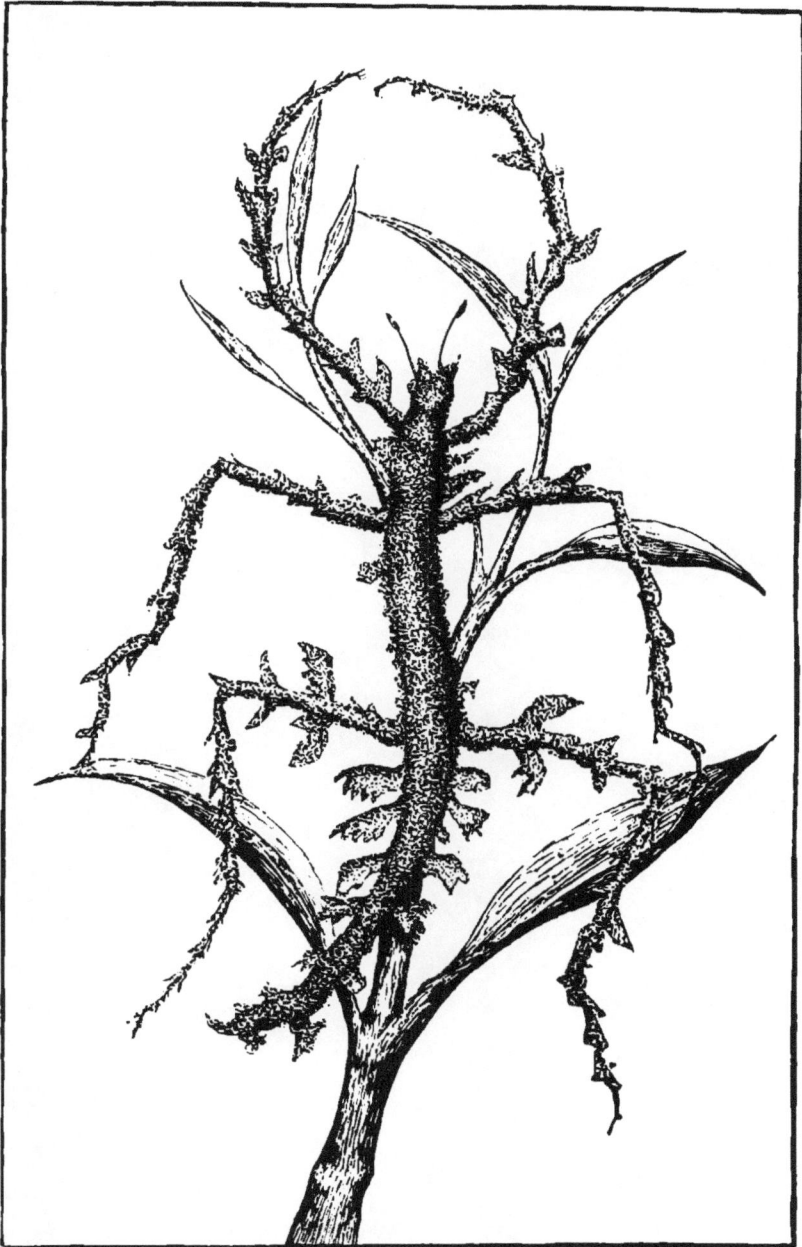

INSECT THAT MIMICS A MOSS-COVERED TWIG.

bird is extinct. Previous to its capture scientists were familiar with it by means of maori traditions referring to a swamp hen that lived at the time of the giant moa and had been a valued article of food to the ancients. In the North Islands it was known as the moho, and in the South as the takahé, and, as it had not been seen since English occupation, it was supposed to be extinct. Mr. Mantell, however, came upon some fishermen who, in searching for seals, had found the track of a strange bird, which they captured after a long chase, in a gully on Resolution Island. It fought violently, uttering loud screams. It was two feet high, and ornamented with rich purple, green, and golden tints. The sealers had eaten the bird, pronouncing it delicious, and to this fact it probably owes its extinction.

In 1796 Ledrus gave a list of fourteen kinds of birds observed by him on the islands of St. Thomas and St. Croix ; and now, eight of these have become extinct.

According to M. Guion, there were at no distant period six species of *Psittaci* on the islands of Guadeloupe and Martinique, which are now exterminated.

The list of animals verging on extinction is very large, and without the gift of prophecy, we may predict that the bison in fifty years, if not swept away, will have become exceedingly rare. At present their range is between the upper Missouri and the Rocky Mountains and from Northern Texas and New Mexico to Great Martin Lake in latitude 64° north, and year by year it is growing smaller and more constricted.

· With them, to the common fate of extermination, the Indians are surely passing,—a forcible example of extinction within the memory of man. Within a few hundred years mighty tribes have been swept away before the advance of civilization. The Eastern States, one of the centres of barbaric power, have but

11

their annals to record the fact.   The powerful tribes of Georgia are represented by rude works of pottery in our halls of science. The Florida braves are rapidly disappearing; the entire race of red men, the former kings of the New World, forming but a pitiable study for the ethnologist of the Nineteenth Century.

CHAPTER XIV.

THE INK-BEARERS.

The cuttle-fish bone that our canary pecks so eagerly, and the sepia of the artist in water color, are some of the practical offerings of the ink-bearers, and when we remember the commercial value of the one, and the importance of the latter in olden times, it will be seen that the cuttle-fishes are not without their value to mankind.

The head-footed animals, or Cephalopods, as they are called, embrace a large and varied assortment of animals known as squids, octopods, argonauts, spirulas, etc., nearly all having a wide geographical range and an extremely ancient ancestry. Fossil remains have been reported where the shells were almost as large as a cartwheel, while the *Orthoceros titan* must, says Dr. Newberry, have weighed some tons.

With such ancestors, we would not be surprised to find modern representatives of huge size, and, within comparatively a few years, this assumption has been fully justified. For many years the various works on natural history have contained vague references to gigantic squids, but not until 1876, or thereabouts, were the tales verified by the capture and exhibition of a large specimen. This individual I examined with Dr. Holder and Professor Verrill, and it was found to measure about forty-five feet in length, and, in appearance, to fully justify its popular name of devil-fish. The body was long, slender, and sack-like, as shown in (*Plate XXV.*), ending in an arrow-shaped tail. The head, with its

163

enormous eyes, seemed separate from the body, and was apparently divided up into ten arms or tentacles, that extended before it, eight being of medium size, while two were extremely long, having a group of suckers at their enlarged end. These arms were apparently used as forceps. They could be joined together or combined in securing prey, and shot out at a victim with great ease and velocity. They are also used to anchor the animals, the latter swinging to them like a ship at her moorings in a gale. The eight short arms had suckers along their entire length, cup-shaped objects with serrated edges, each being a sucker and having a piston-like arrangement, the whole constituting a terrible armament.

Among the tentacles at their base is the mouth, provided with a toothed tongue and a pair of parrot-like beaks of a rich chestnut color, the upper fitting into the lower. Below the mouth is found a tube called the siphon. Water is taken in around the neck to bathe the gills, and by forcing it out of this tube the squid shoots rapidly backward, trailing its long arms or tentacles behind.

When the animal is alarmed, the ink-bag that connects with the siphon, opens its valve, and a cloud of the black fluid is ejected into the water where it becomes quickly diffused, forming an effectual bar to pursuit. The squids also have the faculty of changing their color with great rapidity, and when laboring under great excitement, waves of color seem to pass over them in quick succession. Their motions, as noticed in small specimens, are extremely rapid; darting along with the velocity of light, now rushing into a school of small fry tail first, turning quickly to seize a victim and press it against the bird-like beak where by making triangular nips the vertebrae is generally instantly severed. It is interesting to note that the bite is always in the same place,—the neck.

When darting among a school of young mackerel or sardines, their color is generally pale or light. At other times they may be spotted red or brown, and whenever they settle upon the bottom, as they often do, seemingly to escape observation, they assume the exact hue of their surroundings, and so lie quietly, their presence unsuspected, until a victim passes.

The force with which the squids eject water and ink from their siphons is somewhat remarkable. I have seen a small one hardly six inches long, send a stream into a man's face, who was bending over the water, at a distance of at least two feet, the inquisitive observer suddenly finding himself drenched with ink.

The large squids are found in many waters, though they seem to be especially common on the Newfoundland coast. Until within a few years they have only been known by hearsay, and the disconnected parts found by whalers in the whales. But through the exertions of the Rev. Mr. Harvey, of Newfoundland, some remarkable accounts of them have been reported to the National Museum. The largest, the relative size of which is shown in the accompanying figure, was fifty-five feet in total length, from the extremity of the tail to the ends of the long tentacles. This monster was caught by Stephen Sherring, a fisherman living near Notre Dame Bay. Rowing along he observed the creature partly grounded, and presenting a terrible appearance with its huge glassy eyes. It was churning the water into foam by aid of its enormous arms, and pumping it and ink in large volumes from the funnel, so that the water about it was black. The boatman succeeded in throwing a grapnel into it, the sharp points taking hold, and by fastening the rope against a tree it was securely held, though its struggles were terrific, until the tide went out, when it was left high and dry upon the shore where it was finally converted into dogs' meat.

That these monsters will attack a boat when cornered there seems no reason to doubt. According to Mr. Harvey, two fishermen were pulling across Portugal Cove in 1873, when they observed something floating upon the surface that they took to be wreckage. One of the men threw his gaff into it, when it moved off, proving to be a giant squid. The men stated that its beak was as large as a six gallon keg. It struck the boat, threw out its enormous arms, as if to entwine them, which they severed with an axe as they laid over the boat. Thus wounded the creature darted off, ejecting so much ink that the water was colored for several hundred yards about. This specimen was the *Architeuthis Harveyi*. The one exhibited in New York, distorted with artificial red eyes in the wrong place, was known to science as *Architeuthis princeps*. Within a few years numbers of fine specimens have been found on the Grand Banks floating on the surface, giving rise to the impression that there had been some epidemic among them.

A GIGANTIC OCTOPUS.

Another interesting group of cephalopods is represented by the octopus. They have, as their name implies, only eight arms, and are, as a rule, bottom animals, crawling in and out among the rocks, and drawing themselves through crevices that would seem impossible for so large a body to pass through. They also have ink-bags, and I have often lost specimens by the sudden diffusion of this black cloud, the octopus gliding away through the coral, its rapidly changing hues aiding in its escape. On the Florida reef they rarely exceed two feet across, and were never aggressive, though they were extremely powerful, and as an evidence of their tenacity, I have speared them, and when hauling them in had the animal lift bunches of branch coral weighing from ten to fifteen pounds. At such times waves of color, red and brown, would rapidly pass over the animal ; its eyes would glisten with a green, lambent light, while from the siphon would pour a stream of ink that spread far and near.

The Octopods of the North Pacific are giants compared to these. The *Octopus punctatus*, a form found in Alaska waters, attains a radial spread of nearly thirty feet (*Plate XXVI.*). The body is small, and the tentacles slender and attenuated. Numerous tales are current of their ferocity, but it is doubtful if they willingly attack a human enemy. A case is on record of an Indian girl on the Oregon coast being captured by one, though it should be received with considerable reserve.

Extremely large specimens of the *Octopus punctatus* are often seen in the San Francisco markets, there being eaten by the Chinese, French and Italian inhabitants. The animals are caught by Chinese fishermen, either being speared among the rocks, or trapped in nets of various kinds. Occasionally they are taken in shallow pools where they are left by the tide, and when struck they make desperate efforts to escape, throwing out their long spider-like arms, endeavoring, perhaps, to reach the open water. They are thoroughly marine animals, and never take to the land unless forced.

The octopus though cowardly does not fail to protect itself when attacked, and the sensation of feeling the sucker-lined arms wound about your hands or feet is disagreeable in the extreme, while the pain caused by the serrated edges of the suckers is not inconsiderable.

On the outer Florida reef, they were generally found in the branch-coral, coiled among the branches at their base, and so common that one would be found every ten or fifteen feet in

certain localities. In collecting certain shells, as *Cypreas*, I
would wade along through the lanes that are always found in
the branch-coral beds, and lift great bunches up to examine
them for shells. As soon as the coral was raised from the
water, brittle stars, echini, worms, occasionally an *Astrophyton*
(basket-fish) would drop out in a living shower, and finally a
sprawling octopus, and as I often had to introduce my hand into
the coral to obtain the specimens, stooping over with my head
under water, I would sometimes grasp one of the ugly creatures,
and find my hand for the moment seized in a by no means
pleasant manner.

At times when the coral needed breaking up to secure specimens, a flat boat was loaded and taken to one of the keys, and
the coral thrown upon the beach. On one of these occasions,
one of the men who was assisting a friend of mine, espied a
good-sized octopus crawling from a bunch of coral, some twenty
feet from the water, where it had been placed. The animal
came out with a clumsy motion, and was making its way toward
its native element, its body held aloft and rigid, when the man,
wishing to preserve it as a specimen, placed himself in front and
gave it a violent kick with his bare foot. The next moment he
was dancing to and fro, with the indignant mollusk securely
entwined around his legs, from which position he was finally
released by the laughing observers, who considered it only a fair
return for unnecessary cruelty to the harmless animal.

The octopods range from the tropics to Sitka on the Pacific
side, and quite far north on the Atlantic. A very interesting
form named after Professor Baird, *Octopus Bairdii*, is found off
the coast of Maine, though they are rarely seen in shallow water.
The different members of this class form attractive objects for
study, and one series of experiments will prove of exceptional

interest. This consists in placing several specimens of the same species of octopus in an enclosure in the open water, providing each an apartment with a different colored bottom, and watching the animals adapt their color to it. Some specimens will not change, but in the majority, it is strikingly apparent,—a wonderful example of protective resemblance.

The argonaut is a shelled cephalopod, and occasionally though rarely is cast upon the New Jersey coast. I knew of one specimen that came in at Elberon, and others have been found at Cape Cod.

The spirula is a squid-like form, extremely common on the Florida reef, containing a coiled partitioned shell. I never succeeded in capturing one of these beautiful decapods alive, yet after a storm the keys of the reef, especially Long Key, would be lined with their pearly shells, in every case without the animal, and as they lay upon the beach, the contrast with the myriads of wrecked purple sea-snails, Ianthina, that accompanied them was exceedingly rich.

The latter are much lower in the scale of life, and float about, buoyed up by a bubble float, to which their eggs are attached. When touched they emit a rich purple ink that is almost indelible; at least lasting for many years.

The nautilus has no ink-bag, and is confined to Eastern seas.

We need not be confined to the mollusks in our search for ink-bearers. The spines of the great black echinus that are five or six inches in length, secrete a purple dye, as I have often found to my cost. One of the sea-slugs, a great green creature, commonly known on the reef as the sea-pigeon, when disturbed, emits a cloud of purple ink quite equal to that of a small octopus. In former times the famous Tyrian dye was obtained from certain mollusks, and a large number of animals might be considered as veritable living ink-stands.

CHAPTER XV.

AMONG all families, whether they are human or belong to the lower animal kingdom, we find individuals that from their unusual size are termed giants. From the crabs of our shores it would perhaps be difficult to select a form very remarkable in this respect, although some of the recent deep sea discoveries show some extremely large specimens, while occasionally a gigantic lobster weighing forty-five pounds is caught by the fishermen on our Northern coast. If, however, we go down toward the straits of Magellan, we shall soon find crabs of great size, bulky fellows, with long spider-like legs spreading three or more feet across.

But this South American crab is, after all, but a pigmy when compared to its Oriental cousin, found far away on the coast of Japan. When the first one was seen by a European, it was supposed to be one of the peculiar inventions of the Japanese, and some curious toy or grotesque plaything made in exaggerated imitation of the common rock crab. The traveler who observed them first, saw some claws standing against a fisherman's hut that was situated on the banks of a small river near the sea. The claws were the biting ones, and were long and slender, the nippers seemingly of ivory, so white and glistening were they.

But it was their length that astonished the observer. Each one was ten feet in length, so it was evident that the crab when moving along over the bottom with its claws spread out, as you

170

have probably seen other crabs do, covered an area of twenty-two feet—surely the king of crabs.

The fisherman assured the man from the West that the crabs were common enough in that locality, and later he saw a perfect one brought in by one of the native boats. Its body, or shell, was about twice the size of a man's head, and resembled a rough hewn rock, so that deprived of the long legs, it would easily have escaped detection.

These gigantic crabs were given the name *Inachus*, but it has been changed, and they are now included in a group known as *Macrocheira*. They range from specimens with claws four or five feet in length to twice that. The Cambridge Museum has one of moderate length, and Central Park Museum possesses a body alone that would well cover the head of any of my readers. Such crabs are naturally expensive and a dozen served in this country would cost about twelve hundred dollars, the price perhaps varying with the size.

A gentleman who has visited the localities in which they are found off Japan, told me that they leave the water at night and crawl up the sandy shores of the coast, ostensibly to feed, and when moving along over the beach they present the appearance of enormous spiders. Even when in the water their motions are very slow and deliberate, resembling those of our common lady-crab. The largest of the Japanese crabs was, as given, twenty-two feet across; but the deep sea may hide still larger forms, though it should be remembered that the giants are the exception.

In China crabs are used in medicine, and, curiously enough, a fossil crab ground up into a powder, is esteemed the most, and supposed to have some miraculous virtue.

Among the remarkable giants might be mentioned a strange

crustacean that lived in the earlier days of the earth's history. It differed entirely in appearance from any crab or crustacean now living, having a very long tail, if so it can be called, with many joints, while the head part bore the claws arranged in a very singular manner. These sea-scorpions, for they somewhat resembled these insects in appearance, are now found only in the slabs of limestone in certain geological formations, and *Plate XXVII.*, shows in bas relief, the largest ever discovered, in the slab that was closed about its body as soft mud, the intervening years changing it to hard rock. This *Pterygotus* was about nine feet in length, and surely might be called the king of the crabs of that time. Many of these curious forms are found in the rocks of this country, but none so large as this giant from the Old World.

CHAPTER XVI.

THE TIGERS OF THE SEA.

" In the Louisiana lowlands low."

THE last notes of the old refrain, rendered doubly sweet by negro voices, came drifting over the waters of the outer reef, dying away as our boat crunched into the coral sands of Long Key, which forms a part of the extreme southwestern end of the reefs of the Florida peninsula.

We had "browsed" along, as Scope, our cook, expressed it, down the reef under easy sail, and now, under the friendly gleams of Garden Key light, were on the way to the camp of turtlers and Conchs, whose acquaintance we had made some time before.

Our dingy hauled up on the coral beach, we were soon upon the confines of the camp, which, in the warm, tropical twilight, seemed set for weird and picturesque effects. The moonbeams and the flames from a great brush-fire lighted up the men, boats, and fixtures, casting curious shadows upon the white, sandy beach that stretched away around the curve. The bay was "dead calm," and so still that the far-away "Ha! ha!" of the wakeful laughing gull, and the thundering return of a shark or ray to the water, came distinctly from the outer reef, miles away. Two rude tents, that might have been relics of the Seminole war, were raised against the brush, while several boats, a well-patched seine, and numerous grains and lines, formed the stock in trade. The banjo was never picked by a jollier party, and,

173

lying upon the sand about the fire, they were waiting for the
" 'way-up moon," before hauling the great net. Captain Dave
and myself accepted seats of honor where the smoke was densest,
and the insects consequently least annoying. Scope joined the
musicians to the windward, and the songs rolled on, waking the
dormant echoes of the old reef again and again. Now the rich,
sonorous voices rose in chorus, followed by a laughing solo improvised on the spot, and dealt out lavishly to the inspiriting
picking of the banjo. The stern realities of life had no place
here ; light-hearted, clear of conscience, these boyish men lived a
life of sunshine and enjoyment, and were contented. From far
and near they came, and were at home anywhere. Some sang
from personal experience of the " Louisiana Lowlands Low,"
others of the " Yellow Rose of Texas," and the " Suwanee River,"
while Sandy sang in truth of the " Old Kentucky home so far
away."

The evening was well along when the tide gently surrounded
old Alick's feet, which were extended seaward, a quiet reminder
of work to be done.

"Turn out yere, yo' lazy coons!" shouted the good-hearted
tyrant, who was the recognized boss in perpetuity. "Yo' Sam
an' Pinckney Fust, run in de dingy, an' de res' of we kin ten' de
payin' out. We aint a gwine too have dis yer sene wusted de
way she was over yander. Yo' see, gem'n," he continued,
addressing Captain Dave and myself, who had risen to lend a
hand, " we hauled de oder evenin', an' w'en we swunged in de
net, she bag so, an' I see de mullet beatin'. I give de word, an'
dese yer boys in wif de net wif a rush, an' I'm dogged ef dey
didn't land fifty of de wustest, onaryest mango roots in county
Dade, sah." But the net was in the water, and all hands laid
hold to assist in paying out,—some of the boys wading in with

it, to see that it did not foul. Out it ran like a huge fiery serpent, the meshes, floats, and sinkers waking to life myriads of phosphorescent creatures, that sparkled and glittered like molten gold, and every movement as we waded along threw out streams and flames of dazzling brilliancy that seemed to dart away, veritable creatures of light, into the darkness. When about two hundred feet from the shore, the dingy swept up the beach, Pinckey First (there were four of them) pulling hard and Sam paying out. Finally the end was reached, and they headed in, and when near shore the boys waded out and grasped the line. The silence was broken now: yells, peals of laughter, snatches of song, and heave-hoys rent the air, and under the inspiriting influence of the uproar the net came quickly in, the space between the floats showing decided signs of animation. Here a score of mullets sprang into the air, or some larger fish essayed to cross the line. Myriads of sardines leaped affrighted from the water, the moon-beams glancing from their sides in silvery gleams. In they came with a rush, the finny victims leaping and splashing. The uproar grew fast and furious; everybody shouted and pulled, while old Alick, up to his waist in the jumping mass, encouraged first one side and then the other in inarticulate words and invective, his speech occasionally ending in a hoarse gurgle as he disappeared under water to fish up a mangrove-root and toss it without the magic circle. It was during one of these submarine excursions that he came to grief. The net was well in shore, and nothing was visible of the old man except his bald pate, around which the mullets seemed to play mischievously. It was only for a second, and then up he rose from the sea as if driven from the mouth of a volcano, and with a mighty crash fell upon his face and made for shore, wildly giving orders to drop the net. But it was too late, and, as it came in, the cause of the old man's flight became apparent. A great fish was seen rushing from side to side, confused by the throng of smaller fry and mowing them down with terrific blows. It was a man-eater, and to save the net a sponge-hook was caught in its gills, and after several trials the unwelcome visitor was hauled high and dry upon the shore. The net, for the moment dropped, was now with a rush dragged well upon the beach, and its load of struggling forms hurled upon the sand. How they glistened and gleamed! every tug at the net turning them over in great waves of silvery light, twisting, sliding over one another, the larger tossing the others high in air in desperation, while the patter of the lesser fry was like the falling rain. Mullets with their rounded heads, jacks with golden fins and silver scales dripping with phosphorescent drops, grunts that opened their wide mouths in audible protest, hog-fish, jew-fish, angel-fishes of resplendent vesture, parrot-fishes that vied with their namesakes of the land in gorgeous coloring, snappers red and brown, groupers, sea-shad, porgies, yellowtails, and a host of others, made up this Argus-eyed assemblage, while the crabs, sea-eggs with bristling spines, sea-cucumbers, and other strange creatures that came in entangled in the net would have warmed the heart of my young readers interested in natural history.

The snappers, groupers and porgies were sorted out and tossed into a great car floating near the beach, that was even now overloaded, and would be called for in a few days by a smack in the Havana trade. The mullets were reserved for home consumption, and, finally, the great net was hauled up on the shore to be cleaned for the morrow.

"Is'e been on de back," said old Alick, "of nigh on to eberyting in dis yer country, from a wil' steer toe a manatee, but I never did 'spec' toe be toted by a sherk. He dash right 'twixt my legs,

PLATE XXII.

A CATERPILLAR THAT MIMICS A SHREW.

an' den sen' me blim intoe de air. I don't keer fo' any mo'.
I'se talken now." "He's good for a gallon of oil," spoke up
Sam. "Dat's a fac'," rejoined the old man. "I didn't ride him
fo' nuthin', son; an' I 'spec' we may as well try fo' mo' in de
mawnin'."

The prospect of having a chance of hauling in a man-eater
from the shore was enticing, and we decided to remain on the
beach all night and join in the sport on the morrow.

"We do a right smart business in shark-oil," said Sam, as we
resumed our places around the fire, an Adirondack "smudge"
of bay-cedar. "We try out the oil, an' when we gits a barrel
we ship it up to Key West."

"What is the oil used for?" I asked.

"Well," replied old Alick, with a mysterious air, "dat's
reliably one of de secrets of de trade. Dis yer sherk-oil goes to
Key West an' Jacksonville, dat's sartin sho', an' dey say dere's
a right smart call fo' cod-liber oil on 'count of dese yer inwalids
a-flockin' dere. Jes' were de oil goes I can't say; you kin dra'
yo' own influences."

The "influences" opened a field of speculation too extended
for the lateness of the hour, and soon the sands of the key
resounded with the hoarse breathing of the whole camp.

Our morning toilet consisted in shaking the hermit ör soldier-
crabs from our clothes, followed by a swim in the warm water,
after which we turned our attention to the fried mullets that
Sam was turning with a mangrove branch fork to the tune of
"Ham fat fryin' in de pan."

The boys were soon at work; the trying-pots were taken to a
small inlet lower down the beach, and five stout poles were
driven into the ground, about fifty feet apart, to which the lines
were attached. These latter were ropes about two hundred feet

12

in length. The hook was a gigantic instrument, eight inches across, that worked on a swivel attached to the line by a three-foot chain. The bait, a large grouper, was fastened on, and then the lines were towed out by the dingy, and thrown over fifty feet from the shore, near the channel. Each line at the water's edge passed over a crotched stick, the fall of which was the signal of a bite.

"Did any one ever get bitten by a shark about here?" asked Captain Dave as we lay stretched out in the shade of the bay-cedars, waiting for a nibble.

"I knew of one case," replied Sam, "down by Sea-Horse Key, and the man was my own cousin. We was goin' out the southwest channel, when the sail jibed, and the boom struck Dorsey an' knocked him over. I threw the oar at him, but before I reached him he threw up his arms with a terrible scream, an' went down. It was a great place for tiger sharks, an' one must have taken him. They jump ten feet out of water an' take fish hangin' over the stern of a smack. Off goes my line!" he shouted suddenly, jumping to his feet.

All hands followed suit, and, sure enough, the stick was down, and the line twitching and jerking as if a curious crab was nibbling at the bait. Most sharks bite in this way from the bottom, nosing the bait before starting off.

There had been some dispute the night before among the boys as to Scope's abilities as a shark fisherman, he having claimed that he could catch a shark single-handed; and now, at his request, the rest stood back to receive a few lessons. Scope had been our faithful cook for years on many a trip about the reef, but had never confided to us that he was a shark expert; but under the taunts of the mainland darkies he had rushed recklessly upon his fate. He took the line from Sam's hand just as

it began to run out, and stepping back about five feet from the
water's edge, planted his bare feet in the treacherous sand and
"paid out," while the other boys stood around, loquaciously
questioning as to whether he had a large family to leave and
had made his will. But Scope kept his eye on the line, paying
out gradually as the fish swam off, and finally, when he
thought the bait had been swallowed, he braced back; the line
tautened, grew rigid, and at this supreme moment he gave a
mighty jerk. The result was unexpected. A cloud of sand, a
pair of heels in the air after a black object *in transitu*, a terrific
splash, and a yell of laughter greeted our unfortunate cook as
he picked himself up, ten or twelve feet from where he had
originally stood, and scrambled ashore. The fish had fairly
jerked him into the water. But there was yet a chance to redeem
himself, and, grasping the hissing line, he lay back upon the sand,
only to let go in time to save a repetition of his late experience.

"Haul in de fish," said old Alick, shaking with laughter.
"Yo'se a born sherker, sho' 'nuff."

Scope looked at the rigid rope, and laid hold in desperation.
Suddenly the line slackened, and, with a look of triumph he
threw the line over his shoulder, and started up the beach on a
run. The shark was swimming in, a trick they have, often
breaking a line in the rush that always follows.

"Look out!" cried Sam.

But it was too late. The great fish, still unseen, suddenly
changed its tactics, and a terrific jerk threw Scope backward in
almost a complete somersault, filling his mouth and eyes with
sand and almost breaking his back; before he could recover, the
line, which had a turn around his wrist, nearly dragged him
over, but from this predicament he was rescued by the laughing
boys, and, chafed mentally and physically and thoroughly dis-

gusted, he gave place to Sam. How simple it all seemed—now hauling, slacking out, jerking the line this way and that, running up the beach as the shark made desperate rushes from side to side, Sam was surely the Walton of the shark line. Suddenly he lay back upon the beach, almost prostrate, his feet ploughing deep furrows in the sand, while the man-eater, as if enraged at this resistance, rose fairly in the air above water, and shook its ugly maw in desperation. But the clanking chain was its funeral knell. A few more leaps and surges, and the monster was humbled. "Clap on yere, boys!" cried Sam. All hands quickly seized the rope, and with a rush the shark came in, lashing the water with terrible blows, running its shovel nose into the sand, and was finally landed high and dry, snapping its jaws in savage defiance. It was a noble one—over thirteen feet long. Alick finished it with grim satisfaction, and was preparing to commence a post-mortem in the interests of the Jacksonville invalids, when another line was noticed running out farther down the beach. Wishing to try the sport, Captain Dave and myself started for the rope, followed by Scope, but before we reached it the line came taut with a thud, and the post tore from its bed with a spring, and dashed into the water. Upon the impulse of the moment, we launched the dingy that lay upon the beach, and were soon in full pursuit of the log, which was rushing up the channel at the rate of ten knots an hour. It would have been a fruitless chase, but for the fact that the channel, like many others on the reef, was a blind one, ending in a shallow, coral-lined bank. The fish soon reached the end and turned, and the log came tearing toward us."

"Steady!" shouted Captain Dave.

"Steady!" gasped Scope in a hoarse whisper, himself very unsteady with excitement and his late exercise.

As the post shot by, Captain Dave, who was in the bow, grasped it. The little boat whirled about madly, throwing us down among the oars and bailers, and Scope, utterly demoralized, begged the captain to "cut de rope." Up along the beach we tore, two of the boys putting out to lend a hand; but our steed was only warming up, and the rope they tossed us was missed as we went by. "This won't do," gasped Captain Dave, red in the face from the exertion of trying to keep the line in the notch at the bow; "we're going out the channel."

The boys, who were pulling after us, shouted, "Take in the slack!" This we endeavored to do, but every movement on our part only spurred the shark on to greater feats of speed, and the dingy was now taking everything as it came, and was nearly half-full of water. We hauled away, now gaining a foot, then losing it, when suddenly the line slackened. "He's gone," exclaimed the captain, wiping the salt water from his eyes; "the line's broken." "Bless de Lord!" began Scope—but he went no further.

It was the old trick. The line fairly screeched through the water, slipped from the bow, and in a second was over the side. "Cut it, Scope!" shouted Captain Dave, leaning up to windward. But Scope was getting to windward himself; with a surge the rope caught under the dingy, the opposite rail flew up, and for a single second we were high in air; then with a slash, the water came in upon us, and the boat righted half-full of water, and rushed away in the opposite direction from that we had taken a moment before.

We were soon picked up by the boys, who played the great fish, hauling him in, jerking him this way and that, till finally, when he was thoroughly beaten, the line was passed astern, the monster drawn to the surface, and the chain made fast. Though powerless to swim, its great tail lashed the water into foam, and, after bending its body into a curve, it would suddenly straighten out, lifting the boat out of water. At last, as if in desperation, it seized the keel, crunched the pine planks, and shook the boat as a cat would shake a mouse. But the shark was fairly caught and slowly we towed it ashore. The line was soon passed to those on the beach, and the man-eater run upon the sand.

The sharkers had not been idle: three large fish were thrashing upon the beach. The one that had led us such a wild chase was hoisted upon an improvised derrick, and found to measure fourteen feet in length, while the others were not under twelve. Their girth was so enormous, however, that their great length was hardly appreciable.

The strength of these brutes is surprising, yet by skilful management and an ample allowance of line, a boy can subdue one, though a dozen or so men will be required to land it. Among the many I have caught during a long residence on the reef, a striped shark showed the most power. I had fastened my boat to a clump of branch-coral by a coral-hook that could be readily cast off, when ten or more sharks appeared suddenly, attracted, perhaps, by the bucket of beef blood that my man tipped overboard. I soon had a bite and caught a glimpse of the great fish as he took the bait. After giving him twenty feet I jerked the hook into his jaw, and was almost hauled out of the boat by his answering pull. The coral-hook was taken in, and in a moment we were flying up the channel, bow under, at race-horse speed. I stood on the little deck of my boat with my foot on the line to keep it in the notch, and glancing over the side I saw four or five of the sharks were following along, their sides and eyes turned upward as if in curiosity. I must confess that their attendance gave me a most unpleasant feeling.

For half a mile our steed towed us, then suddenly changed his course, by a miracle not upsetting us. The line, despite our utmost endeavors slipped over the side, and more than once I had the knife over it to cut it, but finally we got it righted and the monster on the back track. A friend on the key watching the struggle had put out in a twelve-oared barge, and as we rushed by they threw us a rope. For a moment the shark held against the twelve rowers, then came an ominous sound, the line broke, and he was off. We could not tell his size, but judging from the amount of strength he displayed, he might have been fifteen feet or over. On all the sharks captured was found the remora, a peculiar sucking fish. Three or four would always cling to their great consort even after it had been hauled upon the sand. Pilot-fishes were also always seen about the sharks here. I have often seen them dart away from their huge consort, and swim swiftly back, as if to carry the news, but this is purely imaginary, as the fishes probably live about the shark as a matter of protection, just as others of the family are found under jelly-fishes, the physalia, etc.

The sharks are the lions, tigers, jaguars, and wild cats of the ocean world; differing in form and methods of life, yet calling to mind these animals in all their habits. So we can compare them to the rapacious birds. The great man-eater is the vulture; the dog-fish resembles the predatory hawk; while the large basking-shark reminds us of the great condor, that, though extremely powerful, often prefers the smallest game, and that at second-hand.

The sharks, of which about two hundred and sixty species are known, differ greatly from the other denizens of the sea, and are literally without bones, the skeleton, if it can be so called, being made up of cartilage. The centre of the vertebral column is only at times more or less bony, and the dorsal cord does not always exist; notwithstanding this, the vertebræ are generally distinctly indicated. The parts of the skull are not united by sutures, as in other fishes. The gills resemble straps, and vary in number. They have no air-bladder, that is such a prominent feature in true fishes. Their eggs, few in number, are often peculiar in form, being inclosed in horny capsules, have four handle-like feelers or tendrils, that have the faculty of grasping sea-weed, or other objects, as soon as deposited, exactly as does the advance tendril of a grape-vine; thus the egg is prevented from washing ashore, and often they so resemble the surrounding weed as to find protection in the mimicry. In many species, the young are born alive.

In general appearance the shark is repulsive. The skin is rough and file-like, being protected with minute or hardened granules; the mouth is generally placed beneath, and armed with sometimes eight or nine rows of sharp, serrated or saw-edged teeth, all except the first having the faculty of lying down, a reserve force that is only used as occasion requires, rising up so many hooks, to secure, hold, and lacerate the struggling prey. They are found in all waters, from Arctic to Equatorial, and in both salt and fresh.

Perhaps no animal excites so much dread as these monsters of the deep. They are the scavengers of the ocean, and though their ferocity is often exaggerated, there are many cases on record showing that they are not feared without cause.

While on the reef, I often in company with others swam out into the channel in the direction of a neighboring key, when sharks ten or twelve feet long had been seen to pass a few moments before. Their presence was not thought of, perhaps because no accident ever occurred, and they were so common;

again, they were well fed from the neighboring slaughter-house. The only time I ever saw any disposition to follow a person was one day I was wading along in water about up to my knees, when a young shark, about four feet long, suddenly appeared and darted after me, whether in fear or play I did not wait to ascertain, having left my grains in the boat. It was nearly two hundred feet to the fringing bare reef, and the way was paved with echini, craw-fish, and young coral. I was barefooted, but I must have astonished the shark by my leaps, as I reached shoal-water well ahead, and pummeled my follower with sea-cucumbers, and short white-spined echini, at which it darted away so rapidly that I was assured that its action was only play.

Sharks occasionally attack boats, either thinking they are fish or with more vicious intent. For several years fishing dories along the Maine coast were attacked from time to time by a huge fish that the men called the great biting-shark. It made desperate attempts to get into boats, and a fisherman at York informed me that his grandfather was attacked by it, and with difficulty drove it off, the boat nearly tipping over. Accounts were heard of the shark for a long time along the coast. A few seasons ago a fisherman was sitting in his dory off the Nubble, near York Beach, when a shark, nine feet long, deliberately jumped into the dory and began hurling the oars about, much to the astonishment of the fisherman. In this instance the fish had evidently made a mistake.

The shark generally known as the man-eater of the American coast is Atwood's shark, a rare and formidable creature, bulky in the extreme, and attaining a length of thirteen feet. It ranges from Newfoundland to Florida. One struck by a harpoon, some time ago off Cape Cod, turned upon the boat, seized the cut-water in its mouth, crushed it badly, and only succumbed when it had been repeatedly struck with the harpoon and lance. The most ferocious of all the sharks belong to this genus. Fortunately in our waters they do not attain their full size, but in the British Museum is a jaw of a man-eater, captured in Australian waters, that measured thirty-six feet and a half in length, more than twice the length of the largest so-called man-eater shark that I have observed on the Florida reef. Such a fish would be a much more terrible enemy to encounter than a whale, and a match for a large boat. The jaw of such an one would well allow a man to stand upright in it, and a man or a horse would be but morsels to appease its appetite.

Among the predatory sharks is the mackerel shark, a beautiful fish, somewhat resembling that after which it is named. They attain a length of ten feet, and a weight of four hundred pounds, and greatly annoy the fishermen by biting off their hooks. The thresher, or fox-shark, attains a length of twenty feet, with a tail of nearly half that length, and presents a remarkable appearance; while others are the blue shovel-nosed, dusky, and hammer-head. The latter has two projections, or lateral, hammer-like prolongations, upon each side of the head, that bear the eyes, giving the fish a most singular appearance. It is much dreaded for its boldness and ferocity.

It is not generally known that sharks live in fresh water; yet such is the case, one kind having been seen in Lake Nicaragua by Belt, the naturalist; and in the River Wai Levi, in Fiji, another shark, the *Carcharias gangeticus*, also lives. This shark is likewise found in the Tigris at Bagdad, three hundred and fifty miles from the sea. In Viti Levi it breeds in a fresh-water lake above the falls, where there is not the slightest possibility of the water being even brackish.

The man-eater, *Carcharias*, is by no means the largest shark.

In the waters about the Seychelles Islands a shark is found that attains a greater size than many whales, some reaching, it is said, seventy feet. The native whalers often mistake them for those cetaceans, and harpoon them, but the error is soon seen, and, it is said, so lightning-like is their rush and dive that the rope is often burned or entangled, and the boat torn to pieces. In one case one of these monsters was struck, and as it dived the men, seeing what it was, rushed to the line to cut it; but this was impossible. The men cried for help, and several dashed into the water and started for another boat, when suddenly the line became exhausted, and with a lurch the *Pirogue* disappeared. Some days later she was found a number of miles away, completely wrecked.

The *Rhinodon typicus*, for this is the name of the monster, occasionally upsets boats by rising under them (*Plate XXVIII.*), evidently taking them for others of its kind, but otherwise than its terrible strength it is perfectly harmless, not having the sharp teeth and ferocious nature of the man-eater.

Comparatively few of these fishes have been caught. One of the finest specimens, now in the Colombo Museum, was captured in the Indian Ocean in 1883. Its length is twenty-three feet; girth, thirteen feet. It was taken in what the Cingalese call a *mahadthalle*, or great net, that they run out nearly a quarter of a mile into the sea. Mr. Ward, friend of Dr. Wright, of the Dublin University, measured one that was forty-nine feet long, and the latter saw others that exceeded fifty feet. He likewise obtained the assurance of competent witnesses that they had been seen seventy feet in length. Dr. Wright verified his statements of their great size by seeing specimens harpooned and photographing them.

This, with the exception, perhaps, of the basking-shark of our shores, is not only the largest living fish, but the largest animal in existence next to the great rorqual-whale.   Like the whales, they prey upon extremely small and pelagic animals, and have certain modifications for this purpose similar to that of whales or the basking-shark.   The mouth, instead of being underneath, opens on a level with the snout, and in it are found cartilaginous bands, analogous to the whalebone of whales, evidently for retaining small prey.

Equaling this monster in size, is the basking-shark of our own waters.   It is known under a number of names, as sunfish, sail-fish, hoek-mar, and bone or elephant-shark.   It is quite harmless, its food being the small animals that float at the surface of the sea ; the mouth having a whale-bone arrangement calling to mind that of the whalebone whale.   On the Greenland coast they are caught in great numbers, the most important fishery being at Naorkanck, where three hundred or more are taken during the season (a short one), their livers yielding about two thousand four hundred barrels of oil, which is preferred to seal-oil, and finds a ready market and good price at different ports in Europe.   The fisheries on the coast now extend beyond Fiskenaes and Proven, where the "spee," or blubber, of "hoewealder," as the Icelanders call the great fish, is taken as a medium of exchange for tobacco, pipes, coffee, and other luxuries from the outer world.

There is a legend recorded by Mitchell that bone-sharks were formerly caught at Provincetown, Cape Cod, in paying quantities.   Twenty years ago one was washed ashore off Rockaway that was thirty feet long.   Earlier than this one came ashore at Cape Cod, of such gigantic proportions that the inhabitants went to the beach for blubber, thinking it a whale. Seven barrels of oil were taken from its liver and sold in Boston for one hun-

PLATE XXIII.

MAMMOTH ADRIFT ON AN ICE FIELD.

dred and four dollars. About the Orkney Islands is a favorite place for them, where they are called the hoemar.

Mr. Daniel Perkins, School Commissioner of Wells, Me., informed me that the Gloucester fishermen claimed the largest specimen of this shark ever caught—a monster seventy feet long. Later, to refresh my memory, I wrote him upon the subject, and the following is his reply :

" Your remembrance of the story was mainly correct. The facts are these : The schooner Virgin, of Gloucester, of which vessel one of my neighbors, now deceased, was one of the crew, caught a shark off Block Island from which they took eight barrels of liver. They lashed its head to the windlass-bitts, and his tail extended past the stern, so that he was longer than the vessel, which was of sixty-eight tons burden. They also struck another shark the same day which they reported larger, but he took their harpoon and line. . . . . Several well-authenticated stories of sharks of nearly equal size are reported. My great-grandfather emptied a pan of coals on the back of a shark which was lying alongside of his vessel, on the Grand Banks, which he said was longer than the vessel."

On the coast of Portugal is a great shark fishery. The fish are brought up from a distance of nearly two-thirds of a mile in the bay, and are invariably dead when they reach the surface.

Strangely enough, the smallest sharks are often the most dreaded. The simple dog-fish has within its power to cut off the entire income of the fisherman. On the Maine coast, during August, where boatloads of cod, hake, and haddock are brought in to-day, to-morrow not one can be had, the cause of this peculiar disappearance being the sudden appearance of immense schools of dog-fish. It would be impossible to estimate the vast numbers of these fish. All other fishing is given up ;

everybody goes dog-fishing, and after several hours the dories come in loaded to the water's edge. The fish are so ravenous that the men spear them while waiting to haul their trawls, and they bite at the sails that drag overboard, and even at the oars and boat. The water seems fairly alive with the starving horde, that bite and devour each other on the hooks. Their livers are sold for the oil, and the skin is often used for the handles of swords, covering boxes, etc.

Concerning the voracity of the dog-fish there is little doubt, and to fall overboard out upon the Banks would often be fatal. The visitation of these fish is of yearly occurrence, generally during the month of August. The fishermen believe them to come from the Gulf Stream and warmer water, as they disappear upon the approach of cold weather.

In English waters they are equally a scourge, and as many as 20,000 have been caught in the single haul of a net. Their numbers are beyond conception. A fair example is a school of picked dog-fish observed by Professor Couch, who states that they extended in an unbroken phalanx from Moray to Aberdeen, and from twenty to thirty miles to sea. They are used as guano, and in many parts of the New England coast large signs are seen, " Dog-fish factories," " Dog-fish bought and sold," etc. Here the fish are taken, ground up, dried, and prepared and sold as soil-dressing, or as a substitute for guano.

The economic value of sharks is not confined to their oil and hides. The negroes of the New Guinea coast eat the flesh after it approaches the " high " state of excellence so esteemed by epicures in hare, venison, etc. Fifty thousand dollars' worth of shark-fins are imported yearly from Calcutta to China, where they are in great demand for soup. On some parts of the African coast the shark is valued as a god, and named the Jou-Jou. Its month is supposed to be the sure and only way to heaven, and three or four times a year a human victim is sacrificed to it. In some of the islands of the Pacific the teeth are greatly regarded as weapons, being bored at their bases and lashed upon swords, daggers and spears, forming terrible arms, the serrated edges lacerating and tearing the flesh. As a protection from these the natives have a regular armor, made of cocoanut fibre, fine examples of which may be seen as well as the weapons, in the archæological collection in the Museum of Natural History, Central Park. The most formidable weapons are a pair of gloves, or long gauntlets, that cover the arms, and are faced with long recurving teeth. These are only worn by the largest men, who, in battle, rush boldly into the throng, seize a victim in their arms, and literally tear him to pieces. The real value of the shark, however, is its work as a scavenger; it, with the vast droves of dog-fish, forming the purifiers of the sea.

The fossil sharks form an interesting chapter, especially from their gigantic size. The man-eaters of the Tertiary time attained one hundred and twenty-five feet in length. . The teeth of these great sharks are very common on the site of Charleston.

Interesting in this connection is the Port Jackson shark, four species of which have been discovered in the Indian Ocean. It has crushers instead of teeth, and is altogether a most remarkable creature. But strangest of all, it seems one of the few forms of the ancient geological times preserved unto this, closely allied species being common in the older geological formations.

CHAPTER XVII.

LIVING LIGHTS.

It is in the Southern seas that the most wondrous displays of phosphorescence are seen. I have drifted over the great coral reefs, where the bottom with its waving plumes and fans seemed studded with gleaming gems, and graceful yellow and purple gorgonias with their reticulated surfaces, were bathed with soft, lambent lights of blue and white, that when taken from the water illuminated all about with a soft radiance. Flashes of light came and disappeared, appearing again in the distant depths like spectres. The silvery sand even turned upon the oar that disturbed it, and gleamed and flashed with sparks of living light. Processions came and went winding away, breaking up and reforming in aggregations of light, nebulæ, of living stars in a watery sky.

On such a night, among the keys of another locality, a party had been floating along in silent admiration of the scenes below, when in one of the boats some distance ahead a singular light suddenly appeared. A large Pyrosoma had been captured, and in its glassy prison, held aloft, in pleasant jest, a living beacon to the more tardy explorers. Although a small one, the brilliancy of this beautiful creature was distinctly visible for quite a distance, so the light of one five feet in height can well be imagined.

The Pyrosoma is scientifically an ascidian, and the branch is remarkable for its phosphorescent peculiarities. Some are stationary, as the sea-squirts, while others as the Pyrosoma,

192

PLATE XXIV.

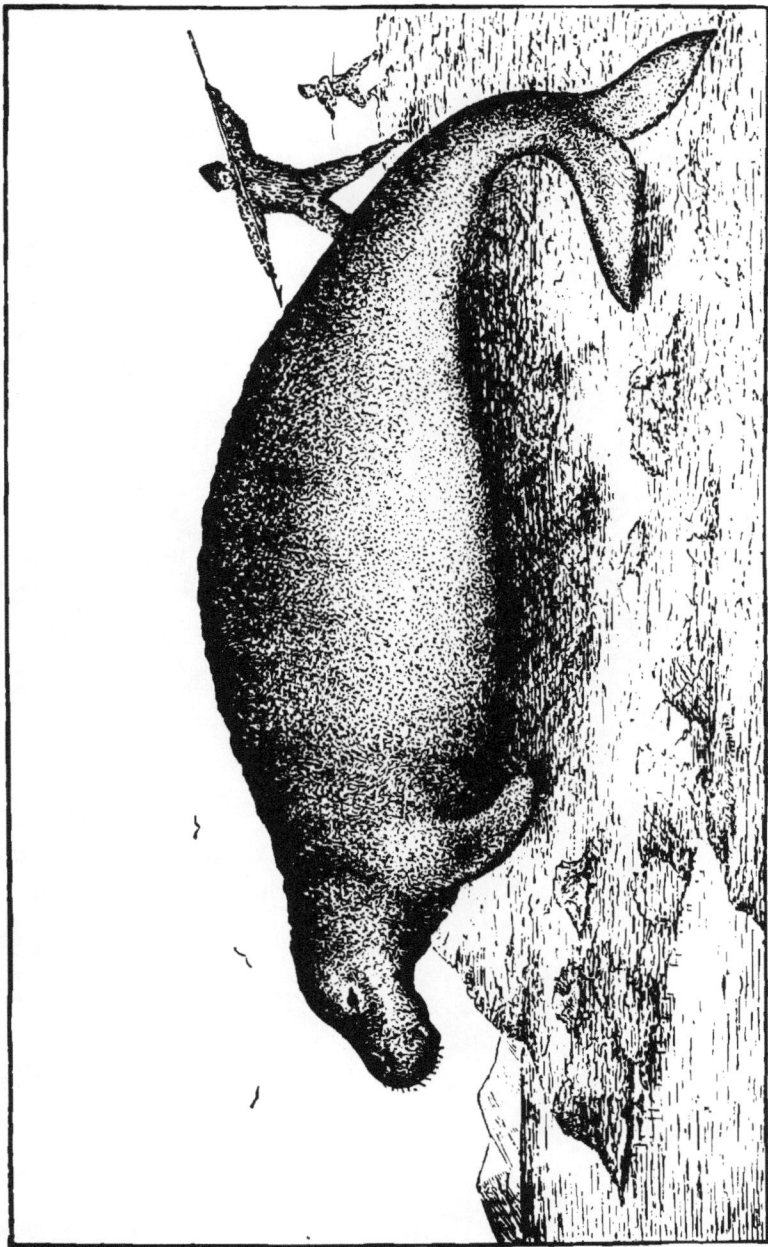

EXTINCT SEA COW (RHYTINA)—THIRTY FEET IN LENGTH.

(*Plate XXIX.*), and Salpa are pelagic. The former is an aggregation of individuals forming a hollow cylinder, closed at one end, and from two inches to five feet in height. The animals, amounting to many thousands, are grouped in whirls, their orifices so arranged that the inhalent are upon the outer side of the cylinder, and the exhalent upon the inner side. Each animal draws in a current from the outside, ejecting it into the interior, and the result of this volume of water rushing from the open end forces the entire colony along. They are richly tinted during the day, but at night, as their name implies, are veritable fire bodies all aflame.

Humboldt refers to the spectacle he enjoyed when passing through a zone of them in the Gulf Stream, distinguishing by their light the forms of dolphins and other fishes, that, bathed by their gleams, stood out in bold relief far below the surface. The light they emit is at times yellow, reddish-green, and azure-blue, and so brilliant that it is said of Bibra, the naturalist, that he used them to illuminate his cabin, writing a description of them by their own light. They are met with in the South American waters in vast shoals; the light of the stars is dimmed, and vessels seem plunging over a sea of molten lava, that breaks into lurid flames at the bow, and clings to the chains in golden drops; while in the depths below, myriads of incandescent forms are seen, the light appearing in flashes that illuminate the ship, sails, and rigging, with an unearthly radiance.

Mr. Bennett, the naturalist, thus describes his experience with these beautiful creatures: "Late one night in June the watch called me to witness a very unusual appearance in the water. This was a broad and extensive sheet of phosphorescence, extending from east to west as far as the eye could reach. I immediately cast the towing-net over the stern of the ship, which soon

13

cleaved through the brilliant mass, the disturbance causing strong flashes of light to be emitted; and the shoal, judging from the time the vessel took in passing through the mass, may have been a mile in length. On taking in the towing-net it was found half-filled with *Pyrosoma Atlanticum*, which shone with a beautiful pale greenish light. After the mass had been passed through by the ship, the light was still seen astern, until it became invisible in the distance, and the ocean became hidden in the darkness as before this took place."

M. Peron observed the beauties of this same animal on his voyage to the Isle of France. The wind was blowing with great violence, the night was dark, and the vessel was making rapid way, when what appeared to be a vast sheet of phosphorus presented itself, floating on the waves, and occupying a great space ahead of the ship. The vessel having passed through this fiery mass, it was discovered that the light was occasioned by animalcules swimming about in the sea at various depths round the ship. Those which were deepest in the water looked like red-hot balls, while those on the surface resembled cylinders of red-hot iron. Some of the latter were caught; and found to vary in size from three to seven inches. All the exterior of the creatures bristled with long, thick tubercles, shining like so many diamonds, and these seemed to be the principal seat of their luminosity. Inside also there appeared to be a multitude of oblong narrow glands, exhibiting a high degree of phosphoric power. The color of these animals when in repose is an opal yellow, mixed with green; but on the slightest movement, the animal exhibits a spontaneous contractile power, and assumes a luminous brilliancy, passing through various shades of deep red, orange, green, and azure blue.

The naturalist Moseley captured a Pyrosoma four feet long, ten inches in diameter, with walls an inch in thickness. It was placed upon the deck of the vessel, and for a long time gave out no light, but writing his name upon the animal with his finger, it came out in letters of fire; each letter then seemed to increase in size, until the entire name was lost in a blaze of light that radiated rapidly, and soon diffused the entire animal, presenting a marvelous spectacle, as if it had suddenly been heated to a white heat, and various chemicals were being thrown upon its surface to produce different colors.

Equally resplendent as a light-giver is the Salpa, a relative of the Pyrosoma. Some species are found upon our own shores, and are sometimes caught in large quantities in Long Island Sound in drag-nets. The curious creatures are free, and habitually swim on the waters of the ocean, and are alternately solitary and aggregated. The solitary individuals resemble short but rather wide tubes, which are often of considerable size, and so transparent that Professor Forbes says they look as if carved in crystal. As they move along on dark nights they often present, when found, the appearance of gigantic fiery serpents, sometimes several yards long, winding their way over the sea—a most awe-inspiring spectacle to witness.

Among the corals the beautiful *Caryophyllia* is an interesting light-giver, and is found in quantities in Northern as well as Southern waters. When the Atlantic cable was taken up for repair, numbers of *Caryophyllia electrica* were found growing upon it in water at a depth of over a mile, and in this spot of intense darkness it perhaps gave out a faint gleam—its contribution to the light of the submarine world.

Professor Moseley, naturalist of the *Challenger*, is of the opinion that the corals are the most important light-givers. All the alcyonarian corals, dredged by the *Challenger* in deep water, were found to be brilliantly phosphorescent when brought to the surface, and their phosphorescence was found to agree in its manner of exhibition with that observed in the case of shallow forms. There seems to be no reason why these animals should not emit light when living in deep water just as do their shallow water relatives.

Some of the sea-anemones, the cousins of the corals, are luminous. The light is generally confined to the tentacles and the smoother soft portion of the column near the summit. We can imagine this beautiful column standing, perhaps, upon some projecting ledge, or at the entrance of some gloomy cavern; its delicate tentacles lighting up, then disappearing as the animal closes, only to reappear cautiously, finally beaming out in all its splendor, in strange analogy to the revolving-light upon the shore above. At times the anemones might be compared to light-ships adrift, as they are apt to take refuge or prominence upon the shell of a gaily-bedecked hermit-crab, that thus travels about, bearing its own lantern, and perhaps preying upon the animals that are attracted by the strange beacon.

In the little boring-shell Pholas we find the most wondrous display of bluish-white phosphorescence. Pliny was the first to notice it, and states that their light was at times so brilliant that the small objects about them were distinctly visible. "Those who eat the *Pholades*," he says, "in an uncooked state (which is by no means rare, for the flavor of the mollusk does not require the aid of cooking to render it palatable), would appear in the dark as if they had swallowed phosphorus; and the fisherman who, in a spirit of economy, supped on this mollusk in the dark, would give to his little ones the spectacle of a fire-eater on a small scale."

Dr. Priestly says this mollusk illuminates the mouth of the person who eats it, and it is remarkable that contrary to the nature of true fishes which give light when they tend to putrescence, this is more luminous the fresher it is; when dried its light will revive on being moistened with salt water or fresh; brandy, however, immediately extinguishes it.

Some very interesting experiments have been made with these light-givers. A single pholas has been found to render seven ounces of milk so luminous that the faces of persons could be distinguished by it, and it looked as if transparent. Many attempts have been made to render the luminosity of the pholas permanent. The best result was obtained by placing the dead mollusk in honey, by which its property of emitting light lasted more than a year; whenever it was plunged into warm water the body of the pholas gave out as much light as ever.

Another beautiful phosphorescent shell is the little pteropod *Cleodora lanceolata*, that, with its distinct head, curious fins, and arrow-shaped shell, swims along, a faint light gleaming through its transparent house, one of the most interesting lanterns of the sea, a pelagic light-ship floating about at its own free will.

CHAPTER XVIII.

WONDERS OF THE AIR.

THE dexterity displayed by various animals in escaping from their enemies may, in many cases, be traced to some special modification of their structure; in other words, nature has provided them with strange and to us wonderful methods of protection. This is shown in an interesting manner in the common flying-fish *Exocetus*, and also in the flying-gurnard. The former is a beautiful silvery, large-eyed creature, resembling somewhat some of the herrings; but as it tosses about upon the deck you immediately perceive that it is quite another fish. The two side or pectoral fins are developed, or enlarged to an enormous extent, so that they are to all intents and purposes wings, and as such are used in a greater or less degree.

The flying-fish is found in many waters, but in the warm currents of the tropics they are seen to the best advantage. In the Gulf Stream I have frequently observed them rise from a wave and dart away over the water seemingly without any perceptible effort, dropping to the water after remaining in the air perhaps for a sixteenth of a mile, and flights have been observed of twice that distance.

When they leave the water, the dolphin or some predatory fish is generally close behind them, though they often leave their native element and soar away in pure amusement. When they rise, often in pairs, or by the score, the tail is seen to work vigorously like a screw, and the wing-like fins to vibrate

PLATE XXV

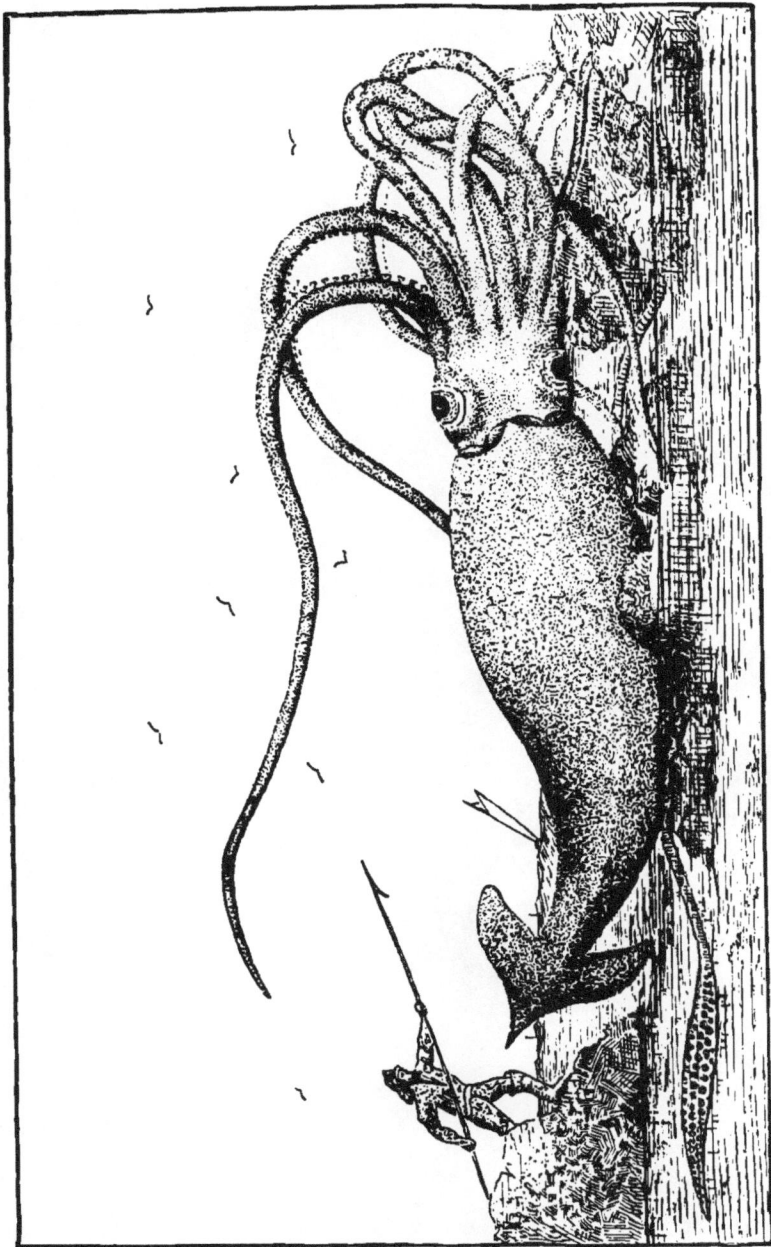

GIANT SQUID—FIFTY-FIVE FEET LONG.

quickly, as if they were exerting all their energy to launch themselves with the greatest force into the other element. After once clear of the water, I have never observed the fins move like wings, though I have carefully watched many hundreds in the air. It is only just to say that some observers claim that they have seen such motion, though I think that examination will fail to show muscular power sufficient to accomplish much in this way.

In the air the fish spreads out the great wing-like pectorals so that they form regular parachutes, and are evidently used as such, as when the impetus is exhausted the fish begins to fall and soon drops. I can only compare the motion to that of the pelican when it rises a few feet, darts down toward the water, and moves along for quite a long distance with its wings out-stretched and immovable, then rising and darting down again to repeat the act that is evidently as much a pleasure to the bird as it is a source of wonder to the spectator.

In performing this feat, the bird skims along the surface within a few inches of the water, just clearing it, and that the wings are held stiff there can be no doubt, as I have observed it not only in hundreds of wild birds, but in a tame pelican that was following me and not four feet from my boat.

When a gale is blowing the flying-fishes often accomplish wonderful flights. As they rise from the crest of a wave the wind catches them, and they are borne away, and often carried as high as the top sail of a ship, striking against the canvas, and falling twisting and flapping to the deck. They have been known to dart through cabin windows, attracted by the light, and a flying gurnard on one occasion struck a sailor upon the forehead in its headlong flight and felled him to the deck.

The gurnards have their heads protected by a heavy thick

armor, and so can strike a powerful blow. Their side or pectoral fins also form great wings or parachutes, and are, as is the entire fish, often very richly colored with red, blue and yellow hues, so that as they dart about amid the sargassum of the Gulf Stream they appear like veritable butterflies of the sea.

The curious little fish *Pegasus volans*, of foreign seas, has web-like pectorals that undoubtedly enable it to leap with greater facility from the water.

Some of the fishes leave the water in long leaps without any special modifications. They are, however, the long and slender forms that seem adapted to such movements.

On the reef the common gars, that are there habitually found on the surface, often leap into the air and dart away, skipping along for some distance. In the East Indies the gars attain a length of several feet, and, according to Professor Moseley, they are often the cause of fatalities from this habit. The natives wade over the reef in search of shells, and sometimes several of these great fishes will rise and skip away blindly, perhaps striking against the wader, and piercing him with the long bill with all the force of an arrow.

A scientist who visits Florida every winter, cruising around the mainland, related to me a curious experience with the pompino, an extremely active fish. The yacht was sailing up a small river, when suddenly a school of fish was seen ahead. A moment later one rose from the water, struck against the sail with telling force, and fell to the deck. Then another arose, and in a few moments the fish were to all intents and purposes bombarding the yacht; some striking, others passing over, clearing fifteen or twenty feet, striking and sliding away over the water a great distance. They came with such force that the men found it convenient to dodge them and get behind the bulwarks.

Among the fliers on land the little tree-toad *Rhamphorhynchus* is perhaps the most remarkable. It was first discovered in Borneo, where its habits might be compared to those of our common flying-squirrel. As shown, the toes of all four feet are webbed to the tips, so that the little creature has four parachutes which it spreads to the wind. Its methods are, as have been suggested, similar to those of the flying-squirrel. It boldly launches itself from high trees, and swoops downward, the broad webs or membranes bearing it up, until it reaches a neighboring branch, when it rises upward a few feet so as to stop its headway, and then easily alights, to climb the tree and repeat the flight, in this way traveling long distances without difficulty.

When in the air the toads present a curious appearance, and sometimes several are observed darting down together (*Plate XXX.*).

Equally interesting are some of the flying lizards, as the draco, that has a membrane between its limbs that also serves as a parachute, by the aid of which they pass readily from tree to tree.

The flying-squirrels, of which there are many varieties in the old world, fly in a similar manner. When the leap is taken, the animal spreads out its limbs to their full extent, the membrane that appears to connect them catching the wind and buoying them up during their flight. In the curious flying lemur there is not only a membrane between the limbs, but also connecting the hind limbs with the tail, thus forming a steering apparatus, so to speak.

A spider, *Attus volans*, common in Australia, has side flaps to its abdomen, which it moves up and down as wings when it leaps from leaf to leaf.

Some spiders adopt a still more remarkable method. They raise the abdomen upward and expel a delicate thread of silk that seems to float upward, growing longer and longer, until finally the wind catches it, and away goes the spider, buoyed up by a balloon of its own making. Sometimes the balloon is a mass of silk caught together, and hundreds of spiders are often seen sailing away through the air in this way.

A GIGANTIC FLYING REPTILE.

In the earlier days of the world's history, the fliers were often of enormous size. The great bat-like creature shown in *Plate XXXI.* was at one time very common on this continent, and was a flying reptile known as the *Pteranodon*. It differed from the European Pterodactyles in being toothless. The figure shown is somewhat conjectural, and is merely intended to give some idea of the size and probable appearance of the animal. From the tip of one wing to another it measured nearly thirty feet, and when a flock of these monsters is imagined in the air, some idea of their appearance may be obtained. They were probably harmless creatures, and judging from the quantities of bones found they must have existed in great numbers. The museum of Yale College possesses the remains of many specimens that varied in size from the above to those not larger than a small bird.

Very much smaller than the Pteranodon was the *Rhamphorhynchus phyllurus*, a curious reptilian creature whose remains have been found in the slate at Solenhofen, Germany. Its jaws were armed with teeth, and its tail was nearly twice the length of the body, long and slender like that of a monkey, or the bat known to science as the *Rhinopoma microphyllum*. But at the end, instead of terminating in a point, it widened out into sail-like semi-oval membranes supported by boom-like bones, the whole arrangement resembling a tennis-racket, and forming the rudder for this strange aerial living craft. Only one specimen has been found, and that is deposited in Yale College. It is one of the most valuable discoveries ever made in this connection from the fact that on the slab the texture of the membrane that constituted the wing is preserved—the first instance on record.

These are but a few of the strange fliers that poised in the air in the olden time, and probably many even stranger forms yet remain to be discovered.

CHAPTER XIX.

ANIMAL TRAPS AND TRAPPERS.

To the lover of nature and her ways, who feels a responsive thrill when meeting her in all her moods, there is vouchsafed an experience and delight hardly to be described. It is a stimulation of the moral and physical senses, bearing rich fruit in its effect upon the mind. Who has not felt the exhilaration that comes with the mornings of early spring? As we walk across the fields to meet the rising sun everywhere new born life greets the eye, asserting itself to all the senses. The fresh green grass bends beneath our feet, the fragrance of the flowers rises—the incense of nature, while the humming of insect life, the carol and song of the birds, with the rustling of the leaves form a harmony of joyous sounds. To the walker, and he or she need not be confined to the woods and fields alone, who thus appreciates the simpler things, those that are too apt to be termed the common-place happenings of nature, there is revealed many phases of life remarkable for their direct and curious analogy with the doings of human existence. In this peaceful corner of the field, or some quiet pool that reflects the sky and trees, the great struggle for existence between various animal forms is in silent progress.

An intimate knowledge of these combatants well repays the close observer. In our own fields examples are innumerable, but perhaps the most remarkable may be found in the warmer regions of the South, where the extremes of cold are not experienced.

204

PLATE XXVI.

GIANT OCTOPUS, PARTLY OUT OF WATER TO SHOW THE SIZE—
TWENTY-EIGHT FEET ACROSS.

In pressing through the vines and wild tangles of the Southern forest in some localities, a common object is a worm-like creature, known as the peripatus, unsightly, perhaps, resembling a lepidopterous caterpillar, of brown or black color, possessing seventeen or more pairs of short comical legs armed with curved claws. The head bears a pair of curious jointed antennæ, evidently sense organs and a pair of simple eyes. In fact there appears to be nothing about this lethargic gloom-loving creature to attract attention ; indeed, it is the simplest of all the insects, occupying the outlying borderland between them and the worms.

As we are observing these details that relegate the peripatus to the common-place, an inquisitive insect approaches, flying about in the uncertain manner of its tribe ; now examining its sluggish and distant kinsman in various positions, again, suspended in the air viewing it from the side, it moves gradually along toward the antennæ of the peripatus that now vibrate gently. Nearer comes the intruder, its wings fanning the other, and finally brushing the antennæ that retract quivering as if with suppressed emotion. The sluggish body that, perhaps, has been taken for a part of the root upon which it has been resting, moves in a convulsive effort ; a myriad of sunbeams flash before it, and swifter than the eye can follow, the luckless inquisitive insect is enveloped not in the jaws of the peripatus, but in a perfect maze of silvery web that holds it firmly in the air. The wings, limbs, and antennæ remain in just the position in which they were a moment ago ; the whole mechanism of life has come to a standstill, involved in a common ruin by this living trap. In a single second the unattractive peripatus has developed a feature of matchless beauty. For several inches about its mouth a cloud of hazy, lace-like fabric seems poised in the air, sunbeams solidified, it might be, for in its entirety it seems composed of slender shafts of glistening crystal, each reflecting the rays of the sun in iridescent splendors. Around about the victim this magic trap has formed, and thus imprisoned the unfortunate insect awaits its fate.

The mechanism of this remarkable trap is not complicated, and can be easily understood by an examination of the mouth of the peripatus. Here we see that the first pair of locomotive organs are turned forward, their claws being modified into a pair of sickle-shaped toothed jaws that grind together in front of the mouth. The second pair of appendages in the embryo in the adult assume the form of short papillæ, called oval papilla, and at their tips small openings are noticed, that lead to large glands where the viscid substance that forms the imprisoning web is secreted. In the glands it is soft and gelatinous, but when expelled, as shown by the muscular action of the animal, it instantaneously crystallizes in the air, forming a perfect net about the victim, constituting, perhaps, the most remarkable defence known in the animal kingdom.

Somewhat similar in its general effect upon the enemy is the attack, or defence, as the case may be, of many of our common land or aquatic planarian worms. In most instances they are inoffensive creatures, incapable of active movement, either aggressive or otherwise, generally finding protection in their resemblance to their surroundings. Let a snail or some enemy touch the sluggish planarian and its antennæ will be seen to withdraw, as if struck by an electric shock. An examination of the skin would show the cause, as it would be found punctured by innumerable spears or arrows, the projectiles of this innocent and helpless appearing trap. These lances are not retractile, and are projectiles in the full sense of the term, resembling in form stiff delicate rods, that are either coiled or irregularly twisted, or even straight, inclosed in cells in the back of the worm. At the moment of attack, as if the creature pulled some nerve trigger, they are released and shot into the air, as if from a catapault, in vast numbers penetrating the offending party, and being, perhaps, fatal to many small and sensitive animals. The effect upon the more delicate portions of the human body, as the tongue, is to produce a scalding sensation, accompanied by a slight swelling. Semper is of the opinion that this method of defence is possessed by the *Onchidium*, a shell-less sea-shore mollusk of the East, that is chiefly remarkable for its number of eyes scattered over the back, and that they are used with effect against its chief enemy, the fish periopthalmus seen in *Plate V.*, that leaves the water, and hops along the shore after it.

In wandering over our fields we shall find in the sandy spots, where the grass has been worn away, a prince of dissemblers, the myrmeleon, and in few insects, and indeed, animals in general, is the presence of seeming thought better shown. Lying upon the ground the actions of the insect may be carefully watched. Curiously enough, it is in the larval stage that its predatory operations are carried on; in after life, when assuming the adult and perfect form, becoming to all intents and purposes a citizen of good repute in the insect world.

Very soon after it hatches, the ant-lion begins its method of action that entitles it to be considered a part, at least, of a living trap. At this time it possesses a long, bottle-shaped body, its small head being provided with an enormously-developed pair of jaws, that appear like gigantic shears. The larva thus armed, and clumsy withal, would present a warning in itself that would prevent its natural prey approaching; so some method has to be devised that will at once conceal it, and at the same time draw without fail the victim within reach of the jaws. With the strategy and skill of an engineer the myrmeleon adopts the pitfall, introducing itself at the bottom as the death-dealing agent.

In forming the pitfall, that is, when finished, a perfect semi-cone, the insect selects a place where the sand is fine, if possible in the grass where small insects are apt to pass, and taking a secure hold upon a central point with its front feet, begins to whirl its abdomen about as a broom or shovel, soon forming a ridge; and by continually altering its position a hollow cone is finally the result.

The ingenuity and patience of these insect workers is often strikingly shown in this labor. Small stones are met with in the excavation, that, despite the whirling process, persist in rolling down into the trap. Innumerable times the myrmeleon will endeavor to hurl them out by the side motion of its body, and only when it has been thoroughly demonstrated to be impossible is another method adopted. The pebble, often twice the weight of the insect, is carefully rolled on to its back, balanced there, and gradually pushed down towards the end of the tail; finally, when in position, by a muscular effort the tail is jerked upward, and the pebble sent flying into the air, often several inches beyond the pitfall. Sometimes the pebble cannot be thrown, and an insect has been seen to renew the attempt over one hundred times before realizing it—certainly a marvelous exhibition of patience. When finally satisfied that the stones or obstruction cannot be removed, the place is deserted and another excavation commenced.

When the pit is complete, it is one or two inches deep, with sides at an angle of about forty-five degrees, the surface covered with fine quicksand that offers no resistance or foothold. At the bottom of this excavation the living trap buries itself, leaving only the scissors-like jaws exposed, and so awaits the course of events. Ants in particular are its victims. Running rapidly along, they are over the edge before aware of it ; the sand so carefully prepared slides them down the incline, and after a few convulsive struggles they are seized by the relentless jaws, and torn in pieces.

The deliberation of the myrmeleon in rending its prey would be terrible in a larger animal, and the *sang froid* with which it places the empty skin upon its back, and shoots it out of the trap, is characteristic of its savage nature. It is often the case that large ants manage to obtain a foothold and escape ; but the ant-lion is prepared for this, and the moment there appears to be a possibility of losing its prey it throws off all concealment, rushes out, and loading its back with sand, hurls charge after charge at the victim, generally causing it to roll demoralized and panic-stricken into the jaws of its assailant.

This predatory life is carried on for two years, when the insect retires to a cocoon, and in a few weeks assumes the general appearance of a dragon-fly. Some of the allies of this insect display equal cunning and ferocity, and all are marked by what in an Australian bushman would be considered intelligence. The Chrysopa deposits its eggs on stalks that are attached to the ground, so that they resemble delicate plants, while the Ascalaphus hedges its eggs in by a perfect fence or paling, that effectually keeps enemies out and prevents the young from straying until they can care for themselves. A larva of an allied form indulges in ghastly humor at the expense of its prey. It destroys great numbers of Aphides, or plant lice, and then attaches the empty skins like scalps to its back, until finally they form a perfect protection, and covering several times the bulk of its

14

body, and in this portable catacomb or chamber of horrors it lives.

In wandering by the sea-shore or floating over the ocean waves, we find many curious and interesting instances of this phase in the struggle for existence. Here is the uncanny basket-fish with its multitudinous bifurcating arms, that are seemingly prepared to drop net-like over some unsuspecting victim. In deeper water we meet with fishes adapted in a marvelous manner to obtain their food with a minimum amount of exertion. One form recently discovered, not only resembles the bottom in color, but the outline of the body appears to be cut up into myriads of frills, cut artificially to resemble sea-weed; and not only are they found upon the sides but upon various parts of the body, covering the fins, and hanging pendent from the enormous and cavernous mouth, so that the entire fish fully and completely mimics the weeds among which it lies.

This resemblance, however, is but part of the trap accessories, as the fish possesses a complete and effective arrangement for enticing its prey near its mouth. Like the human fisherman it is provided with rod and fly, and not one only but several, comparable in their size and shape to trout, bass, and salmon rods. These rods are from six to eight inches in length, and situated upon the neck just in front of the dorsal fin, seemingly modifications of fin-rays. The first, or one nearest the mouth, is of the most importance, and is the largest, generally about the size of an ordinary steel darning-needle, covered with a soft delicate membrane, often marbled or colored to render it attractive. Upon the tip the fleshy lure enlarges, widens out, so to speak, and dangles loosely in the water, and being many colored forms a tempting bait. If rigid the rod would scarcely be of practical use; and here we notice a remarkable mechanism by which the

PLATE XXVII.

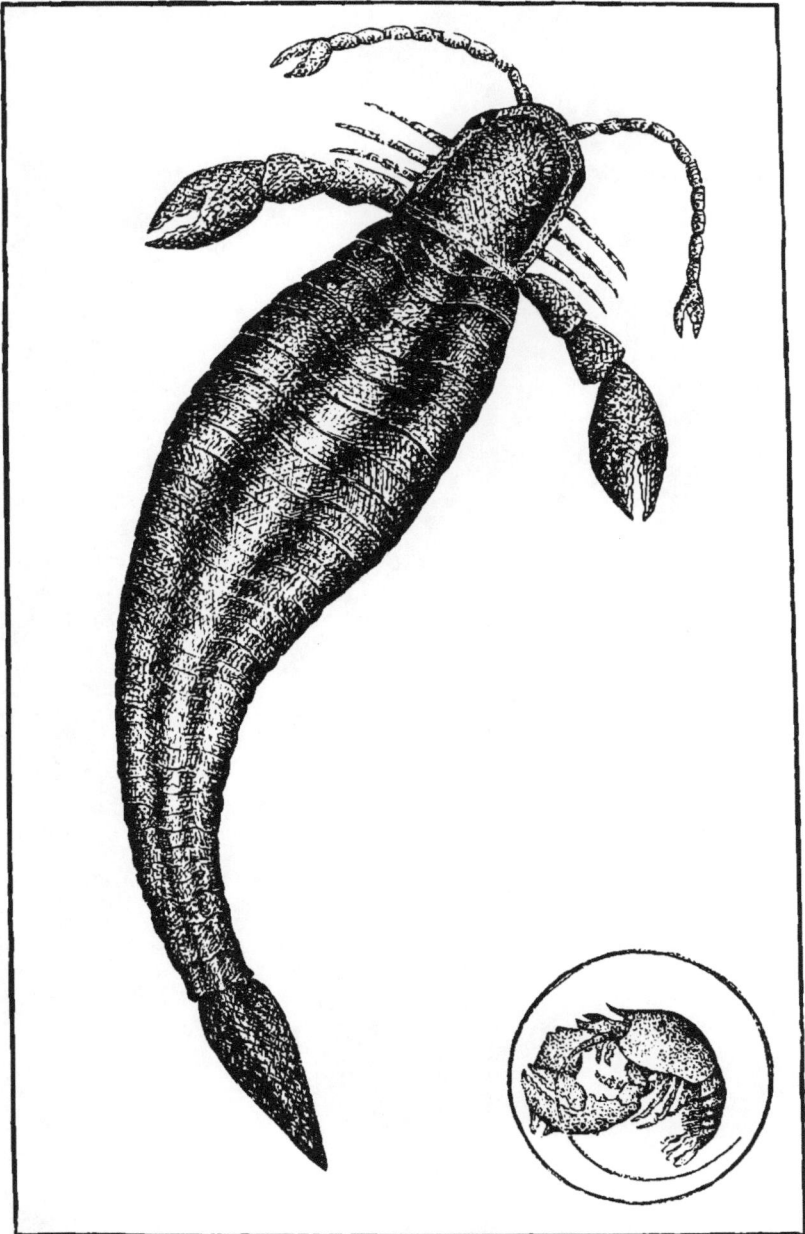

GIGANTIC EXTINCT PTERYGOTUS COMPARED WITH A LOBSTER.

fish can cast its living fly with a mathematical exactness. The rod, or ray, as we may term it, is not connected to the median bones or imbedded in the muscles, but forms at its base a staple that is hooked into a corresponding staple below it so that it works forward and back with perfect ease. The spines behind this act in a somewhat different way, but in the living fish the first one moves up and down at will, and is used exactly as a fisherman uses a baited rod, as a lure. Safe in its disguise of waving barbels the fish flattens itself among the weeds, and lowering its rod casts the filamentous fly exactly in front of the mouth. As some inquisitive small fry approaches, the rod is gently raised, until the victim is beguiled into the right position. This accomplished the abyssal mouth yawns beneath, and into the vacuum the fish is drawn to be lacerated by the rows of movable teeth.

These fishes, *Lophidæ*, do not always depend upon the rod and fly; they have been known to attempt the capture of large birds, and one was secured not long ago, that was found floundering at the surface with a large loon in its capacious maw.

Equally curious is the arrangement found upon the head of the bat-fish, or *Malthea*. It is a soft fleshy object, susceptible of being lowered and raised at will, seemingly a bait without the rod, and supposed to be used as a lure to this living trap.

Some of the fishes dredged from great depths by the late French and Italian expeditions, might be compared to nets, the mouth being the principal part of the body. This is particularly noticeable in the deep sea fish known as *Eurypharynx pelecanoides*. The body is eel-like, destitute of fins, and seems to form merely the handle to the enormous scoop represented by the mouth. So huge is the latter that the fish might well lie in wait, and allow its legitimate prey to wander in out of sheer curiosity. Their method of feeding probably is to lower the under jaw, and use it as a scoop or net, and by pushing along fill it with material of any kind, in some way sifting out the choice bits.

Another form, recognized by naturalists as the *Melanocetus Johnsoni*, seems to combine the qualities and methods of the last two mentioned fishes. The mouth is also so enormous that it can swallow with ease fishes of twice its actual bulk, the projecting pouch below being three or four times the size of the body. The lower jaw, when the mouth is open, extends outward like a platform, armed with an array of teeth equally to be dreaded. Directly over the upper jaw on the median line is seen a long slender rod with a tentacular tip, that constitutes the rod and bait.

As in the genus *Lophius*, with its curious limb-like fins, the *Melanocetus* probably buries itself partly in the mud, and with wide-open mouth dangles the bait, snapping up the victims as they enter the trap so deftly arranged for their reception. To enable many of these traps to swallow prey, there is an arrangement not unlike that found in the snakes, the jaws having great extensibility, and in the *Chiasmodus*, a deep sea carnivorous fish, working independently, actually hauling the victim into the stomach by alternate movements, as in the reptiles referred to.

In China, Siam, and various other countries of the East, the natives train certain varieties of fishes to fight, and the engagements are attended with as much excitement as are the cock fights of the adjoining Malay Peninsula. The fishes selected for this questionable sport are of great variety, and one species comes under the head of our title, possessing a wonderful arrangement for securing its prey only to be compared to the method observed among the native bird of paradise hunters, who use a long blow-gun with which they deftly bring down their game unharmed.

The fish known popularly as the long-billed Chaetodon has been provided by nature with a similar contrivance, the mouth being extended into a long tube—the blow-gun. The projectile is a drop of water, and so accurate is the aim of this finny hunter, that at a distance of two feet, it will strike a fly held in the hand of its owner.

They are marine fishes, frequenting the shores that are over-hung with reeds and grasses, and by swimming along near the surface, the presence of an insect is soon detected, when the beak is extended above the water and a liquid bullet projected that brings the luckless insect quickly to the surface to be devoured.

The Chaetodons are among the most beautiful of the tribe, being decked with gorgeous tints, and for this reason their allies of tropical America are called angel-fishes.

Equally interesting is the Toxotes, or archer-fish, that captures its prey in the same manner, though not possessing the tubular mouth that distinguishes the former. Here the lower jaw is very prominent, extending out like a platform, and evidently acting as a guide to the missile.

In the same waters is found a fish whose mouth constitutes a trap of a totally different character. In appearance the Sly Epibulus resembles somewhat our hog-fish of the South, seeming at a casual glance to be a sober-sided, guileless fellow, incapable of double dealing. This demeanor, however, is only preserved when the fish is surfeited. When hungry it swims slowly and innocently along, causing no commotion among the smaller fry, who, from the extremely cunning way in which their numbers are depleted, do not seem to notice the fact. If the fish is closely watched, a method of procedure will be observed that can only be compared to the game called snatching. The fish moves slowly toward a victim, and when within a short distance away the entire mouth seems to shoot out independently of the fish and grasp the prey, the trap returning again so quickly that unless the action is expected, it is not observed. In this way the Sly Epibulus feeds without approaching near enough to cause alarm by its superior size.

The mechanism of this trap is exceedingly simple. The mouth may be said to be telescopic, capable of several inches' protrusion, and naturally returning to its place, the parts fitting so closely together that such an arrangement would never be suspected.

PLATE XXVIII.

SPOTTED SHARK (RHINODON)—SEVENTY FEET LONG—RISING UNDER A CANOE.

CHAPTER XX.

THE WHITE WHALERS.

"Down with her! Hard!" came hoarsely through the mist.

An oil-skinned figure threw himself heavily upon the oar; the little craft rounded tremblingly up into the wind, hurling clouds of spray and foam aloft that were borne far away by the whistling breeze. For a moment the sail beat furiously, as if in protest at this infringement upon its privileges, then a second oil-skin—the cause of this commotion—raised his arms, a steel spear flashed, a willowy pole trembled in the air, a quick movement, a roar of rushing waters, a shower of spray that drenched the craft, a sound of escaping steam or hissing rope, and a white whale had been struck by Captain Sol Gillis, of Ric.

Captain Gillis, as might be assumed, was not a native of the province of Quebec, but merely a carpet-bagger, who moved North in the summer and returned in early autumn about the time the wild geese went South, and all for reasons only known to himself. He hailed from down East, and voted in a small town not many miles from the historic shell-heaps and the ancient city of Pemaquid.

Our meeting with the down-East skipper was entirely one of accident. Wandering along the beach at Ric, we had come upon a boat, half dory, half nondescript, which from the possession of certain peculiarities was claimed by one of the party to be of Maine origin, and, to settle the dispute, a little house a few hundred yards higher up was visited.

215

It was like many others along shore, single storied, painted white, with green blinds, with a small garden in the rear, in which grew old-fashioned flowers and an abundance of " yarbs " that bespoke a mistress of Thompsonian leanings. A stack of oars, seine-sticks, and harpoon-handles leaned against the roof; gill-nets festooned the little piazza, while a great iron caldron, that had evidently done service on a New Bedford whaler, had been utilized by the good housewife to capture the rain-water from the shingled roof.

"Mornin' to ye, gentlemen. Been lookin' at the bot?" queried a tall, thin, red-faced man, with an unusually jolly expression, stepping out from a shed.

"Yes, we thought she was of Maine build," replied the disputant.

"Wall, so she is," said the mariner, "so she is; and there ain't none like her within forty mile of Rie. I'm of Maine build myself," he added, "but I ain't owner. I'm sorter second mate to Sol Gillis; sailed with him forty year come Christmas. Don't ye know him? What! don't know Sol Gillis?" and a look of incredulity crept into the old man's eye. "Why, I thought Sol was knowed from Rie to Boothbay all along shore. But come in, do. I know you're parched," continued the friend of the skipper, dropping his palm and needle and motioning the visitors toward the little sitting-room. "Mother," he called, "here's some folks from daown aour way."

As the old man spoke, a large-framed woman appeared in the door-way, holding on to the sides for support, and bade us welcome. Her eyes were turned upward, and had a far-away look, as if from long habit of gazing out to sea, but, as we drew nearer, we saw that she was blind.

Leading the way into the kitchen, which was resplendent

with shining pans and a glistening stove, all the work of the
thrifty but blind housewife, she began to entertain us in her
simple manner, and described a model of a full-rigged ship that
rested on a table, though she had never seen it, with an exactness that would have done credit to many a sailor; even the
ropes and rigging were pointed out, and all their uses dwelt upon
with a tenderness strangely foreign to the subject.

"And Captain Sam built it?" we asked.

"No, no," replied the old lady, turning her head to hide a
tear that stole from the sightless eyes. "It's all we've got to
remember aour boy John. He built her and rigged her. He
was his mother's boy, but—"

"He went daown on the Grand Banks in the gale of '75!"
broke in her husband, hoarsely.

"Yes," continued the wife, "me and Sam's all alone. It's
all we've got, and Sam brings it up every summer as sorter
company like. Ye're friends of Captain Sol, I guess?" she
said, brightening up after a minute. "No?" and she looked in
the direction of the captain, as if for a solution of the mystery.
"Naow ye don't tell me that ye ain't acquainted with Captain
Sol, and ye're from aour way, too? Why," she continued,
earnestly, "Sol's been hog-reeve in aour taown ten years runnin";
and as for selec'-man, he'll die in office. Positions of trust come
just as natural to him as reefin' in a gale of wind. Him and
my man tuck to one another from the first."

"Then you were not townsmen always," we suggested.

"No, we wa'n't," was the reply. "My man and Sol met
under kinder unusual circumstances. Tell 'em haow it was,
Sam."

The old sailor was sitting on the wood-box, shaping a thole-
pin from a piece of white pine, and, thus addressed, looked up

with a blank expression, as if he had been on a long search for ideas and had returned without them.

" He gits wanderin' in his ideas when he sots his mind on the '75 gale," whispered the old lady. " Tell 'em abaout yer meetin' Captain Sol, Sam," she repeated.

" Me and Sol met kinder curious," began the captain. " That year I was first mate of the Marthy Dutton, of Kennebec; and on this identical v'yage we was baonnd daown along with a load of coal. In them days three was a full-handed crew for a fore-an'-after, and that's all we had, captain, mate, and cook, and a dog and cat. One evenin'—I guess we was ten miles to the south-'ard of Boon Island—it was my trick at the wheel, and all hands had turned in. It was blowin' fresh from the east'ard, and I had everything on her I could git. I guess it was nigh on ter two o'clock, and as clear as it is to-day, when the fust thing I knowed the schooner was on her beam ends. She gave a kind of groan like, pitched for'ard, and daown she went, takin everything with her; and afore I knowed what was the matter, I found myself floatin' ten miles from shore. I see it was no use, but I thought I'd make a break for it, so I got off my boots and ile-skins in the water, and struck aout for shore, that I could see every once in a while on a rise.

" Wall, to make a long story short, I guess I was in the water a matter o' four hours, when I see the lights of a schooner comin' daown on me. I hailed, and she heard me, ran up in the wind, put aout a bot, and Sol Gillis, the skipper, yanked me in. I couldn't have held aout ten minutes longer. So Sol and me has been tol'able thick ever since."

" Here he comes naow," said the matron, whose quick ear had caught the sound of approaching footsteps. " Sam, set aout my pennyroyal, will ye? Ye see," she added apologetically, " Sol is literary, and when he comes raound he gives us all the news, and there is sech goin's on in the papers now-a-days that it jest upsots my nerves to hear him and Sam talkin' 'em over. Sech murders, riots, wrackin' and killin' of folks! If it wan't for a dish of tea I 'low I couldn't hear to it." And the good woman held out her hand to a burly fisherman in a full suit of oil-skins, and presented him to the visitors as Sam's friend, Captain Sol Gillis.

" I'm a white-whaler at present, gentlemen," said the captain, with a hearty laugh that was so contagious that all hands joined in, scarcely knowing why.

He was a tall robust specimen of a down-Easter, his open face reddened by long battling with wind and weather, and shaved close except beneath the chin, from which depended an enormous beard that served as a scarf in winter and even now was tucked in his jacket.

" It's a curious thing, naow, for the captain and mate of a coaster to be in furrin parts a-whalin', but we find it pays, eh, Sam?" And Captain Sol closed one eye and looked wisely for a second at his friend, upon which the two broke into hearty laughter that had a ring of smuggled brandy and kerosene in it, though perhaps it was only a ring, after all. " Kin yaou go a-whalin'?" said the captain in reply to a question of one of the visitors. " Why, sartin, white whalin's gittin' fashionable. There's heaps o' chaps come daown here from Montreal and Quebec and want to go aout, so I take 'em. Some shoots, and some harpoons, and about the only thing I've seen 'em ketch yet is a bad cold; but there's excitement in it—heaps of it; ain't there, Sam?"

" I ain't denyin' of it," replied the latter. " What's sport for some is hard work for others. Work I calls it."

"Wall, as I say," continued the skipper, "white whalin's gittin' fashionable, so in course there ain't no hard work about it; and if yaou will go, why, I'm goin' aout now, me and Sam. The only thing, it's dampish like; but perhaps mother here kin rig yaou aout."

Half an hour later the two landsmen were metamorphosed into very respectable whalers, and, with the two captains, were running the whale-boat down the sands of Ric into the dark waters of the St. Lawrence. The light sail was set, and soon we were bounding away in the direction of Mille Vaches, Captain Sam at the oar that constituted the helm, and Captain Sol in the bow, with harpoon at hand, ready for the appearance of game.

The white whale, or Beluga, is extremely common at the mouth of the St. Lawrence, and is found a considerable distance up the river beyond Tadousac. The oil is in constant demand for delicate machinery, and Beluga leather, made from the tanned hide, is manufactured into a great variety of articles of necessity and luxury.

In appearance these whales are the most attractive of all the cetaceans. They are rarely over twenty feet in length, more commonly fifteen, of a pure creamy color, sometimes shaded with a blue tint, but in the dark water they appear perfectly white, perhaps by contrast, and seem the very ghosts of whales, darting about, or rising suddenly, showing only the rounded, dome-shaped head.

The Beluga is a toothed whale, in contradistinction to those that are supplied with the whalebone arrangement that characterizes the right-whales; consequently its food consists of fish, and perhaps squid. To enable it to capture such prey it must be endowed with remarkable powers of speed. The motor is the

PLATE XXX.

GROUP OF FLYING TREE TOADS.

great horizontal tail, powerful strokes of which force the animal through the water and enable it to leap high into the air in its gambols. The flippers are small and of little use in swimming. The head is the most remarkable feature. It is the only instance in this group of animals where this organ appears at all distinct from the body. By viewing the creature in profile, a suggestion of neck may be seen, and it is claimed that there is more or less lateral motion—that the head can be moved from side to side to a limited extent. The outlines of the face are shapely, the forehead rising in a dome-like projection and rounding off in graceful lines, so that the head resembles to some extent that of our Atlantic Right Whale (*Balæna cisarctica*).

In their movements the Belugas are remarkably active, and are very playful, leaping into the air in their love-antics, rolling over and over, chasing each other, and displaying in many ways their wonderful agility. They often follow vessels in schools of forty and fifty, and old whalers claim that they utter a whistling sound that can be heard distinctly above the water. The young, sometimes two, but generally one, are at first brown in color, later assuming a leaden hue, then becoming mottled, and finally attaining the cream-white tint of the adult. The calves are frequently seen nursing, the mother lying upon the surface and rolling gently.

The Beluga has a wide geographical range, being found upon our Northern and Northwestern shores in great numbers. Their southern limit seems to be the St. Lawrence, and in search of food they venture up this river beyond the mouth of the Saguenay, and often in water but little over their own depth. On the western coast they also enter the great rivers, and have been captured up the Yukon seven hundred miles from its mouth. In their columnar movements they somewhat resemble the porpoise, long processions being frequently seen, composed of three in a row, perhaps led by a single whale.

Among the Samoyeds, at Chabanova, on the Siberian coast, the white-whale fisheries amount to fifteen hundred or two thousand pood of train-oil a year. On the coasts of Nova Zembla and Spitzbergen they are captured by enormous nets made of very stout material; and the Tromsoe vessels alone have taken in a single season over two thousand one hundred and sixty-seven white whales, valued at about thirty thousand dollars. Magdalena Bay is a favorite place for them, and often three hundred are taken at a single haul in the powerful nets. Here, and in most of the Northern localities, the entire body is utilized, the carcass being used in the manufacture of guano. So perfectly are the bodies preserved by the cold of these Northern regions that if they cannot be removed at the time of capture, they are secured in the ensuing season.

As the boat reached mid-stream, where the wind was blowing against the current, great rollers were met with, that tossed the light craft about like a ball. But this was evidently the play-ground of the Beluga, and dead ahead the white forms were seen darting about in the inky water with startling distinctness, while faint puffs were occasionally borne down by the wind.

Gradually we neared them, and suddenly a white dome appeared on the weather bow. Then came the command and ensuing scene chronicled at the commencement of this chapter.

We were perfectly familiar with whaling-terms, and as the game was struck we construed Captain Sam's impression " git aft" to mean "starn all," and even in that moment of stumbling and drenching felt a sense of disappointment in the suppression of a time-honored term. To omit "There she blows!" was enough, but to substitute "git aft" for "starn all" was a libel on the chroniclers of the " Whalers' Own Book."

There was little time, however, for regrets. Our combined weight had raised the bow a trifle, yet not enough to prevent the sea from coming in ; and, as the skipper, who was laboring with the steering oar, said, the small whaler was "hoopin' along, takin' everything as it came, and askin' no questions." Now by the slight slacking of the line we were high on a wave, the crest of which was dashed in our faces in the mad race ; now down in a hollow, taking the next sea bodily and plunging through it, causing the oars and harpoons to rattle as if they were the very bones of the boat shaking in fear and terror.

In a short time she was a third full of water, and the amateur whalers were invited to man the pumps—namely, two tin basins—and bale the St. Lawrence out as fast as it came in. The maddened animal soon carried us beyond the area of heavy seas, and preparations were made for taking in the slack. The boat was still rushing along at a five-knot rate ; and, as the whale showed no signs of weakening, it was Captain Sam's opinion that nothing short of the lance would stop him.

"Jest lay holt of the line, will ye?" sang out Captain Sol, passing the slack aft, and four pairs of arms hauled the boat nearer the game, that was far ahead. At first this only spurred the creature to further endeavors ; but the steady pull soon told, and after a great amount of labor the white head of the Beluga came in sight.

"Stiddy, naow!" shouted Captain Sol, releasing his hold, and picking up the lance. " Naow, then, work her ahead."

A final haul, and the boat was fairly alongside of the fleeing animal, careening violently under its rapid rushes ; and, in response to the order " Git to wind'ard," we sprang to the weather rail. A moment of suspense, a quick motion of the lance, and the great white body of the whale rose from the water and fell heavily back, beating it into foam in its convulsive struggles.

"She dies hard," said Captain Sol, shaking the water from the creases of his oil-skin as the boat rounded to at a safe distance from the dying whale. "But," he continued, lighting a match by biting the sulphur, and puffing violently at a short, black pipe, "that ain't nothing to what they do sometimes; is it, Sam?"

"I ain't denyin' of it," was the reply of that individual, who was now sculling the boat about the whale in a great circle.

"I've seen," continued the skipper, "a white whale smash a bot so clean that ye'd 'a thought it hed been through a mill; and it was a caution haow we didn't go with it. That was a curious year," he added. "Something happened to drive the whales up here so thick that the hull river was alive with 'em, and of course we was for reapin' the harvest. When we struck the rip-rap, as they call the tide agin' the wind, it was jest alive with 'em, puffin' and snortin' on all sides. I had three harpoons aboard, besides a rifle, and in a minute I had two foul, with buoys after 'em, and as one big feller came up alongside to blow I let him have it with the rifle.

"Naow," he went on, "whether they heard it or not I can't say, but I heard a yell from Sam jest in time to look and see a whale rise I'll 'low twenty foot clean aout of the water. Then there was a kind of a rush, and Sam and me went daown, and when we riz it was gone. The critter had hopped clean over that bot as slick as nothing. That kinder tuck the peartness aout of us, so to speak; but later in the day I got aout the gun agin, havin' broke the lance, and in killin' the critter she jumped

PLATE XXXI.

PTERANODON, AN EXTINCT FLYING REPTILE.—(SPREAD OF WINGS, TWENTY-TWO FEET).

on the bot, and—wall, Sam and me lit aout, and was picked up arter a spell; but that bot, there wasn't enough of her to make kindlin'-wood of.

"They ain't vicious like," continued the skipper, "but clumsy, and if ye get in the way ye're baound to git hurted. Raound the bend at Ric Island one came ashore one time and got left every tide, so she was out of water an hour or so every day. Heaps of city folks went to see her, and one chap came along and let on haow she couldn't live aout of water, and poked her like with a stick. Wall, it ain't for me to say haow many feet she knocked him, but when she fetched him with her flukes it was a Tuesday, and I guess he thought he'd reached the turnin'-pint of Friday when he hauled himself aout of the mud.

"No, they won't exactly live aout of water, but they'll stand it a like of three weeks if yaou splash 'em every hour or so. They sent one to England that way. They ain't fish. Whale's milk's good, if cream is.

"But the best bit of whalin'," continued the communicative Captain Sol presently, "that I ever see in these 'ere parts was done by that identical old chap in the starn there."

"When Sol ain't talkin', gentlemen," retorted the person thus alluded to, "ye'll know he's sick."

"Wall," said Captain Sol, laughing, "I'll spin the yarn, and yaou kin go back on it if yaou kin. As I was sayin', we was aout one day I think a couple of miles below Barnaby Island. I was a mummin' for'ard, kinder sleepin' on and by, and Sam at the helm, when we see a bot a slidin' into the ripple right ahead of us, and in a minute a couple of white heads was dodgin' up a little to the wind'ard. Sam trimmed the sheet and hauled the Howlin' Mary—that's what we called the bot—on the wind, and the other bot did the same, both of us makin' for the same spot.

15

I see it was nip and tuck ; and, knowin' that Sam was a master-hand, I says, 'Sam, yaou take the iron.' So we shifted.

"The other bot had a trifle the weather-gage on us, but both of us, mind ye, makin' for where we thought the critter was comin' up to blow, and in a minute, sure enough, up it come. This 'ere other bot shot right across aour bows ; but, Lord bless ye, it would take a proper good Injun to beat Sam, for he up, hauls back, and let fly the harpoon clean over the other bot, takin' the critter right alongside the blow-hole so neat that the line fell across the other bot—naow, deny it if yaou kin," said Captain Sol, turning to his friend.

"Yer'e a master hand at talkin'," retorted Captain Sam. "I ain't denyin' of it ; but it was luck, good luck, that's all."

By this time the white whale had succumbed, and lay upon the surface motionless and dead ; and upon the boat being hauled alongside the huge creature was taken in tow and soon stranded upon the beach, where the valuable parts were secured—the liver and blubber for the oil, and the thick, white skin that was to be tanned and made into leather or used in the manufacture of various articles.

The evening following, upon invitation, we visited the cabin of Captain Sol, who was a widower and kept bachelor's hall, so to speak. We found him seated on a keg, by the side of an enormous caldron that might have contained the witches compound, judging from the strange forms of steam that arose from it, while the lurid flames beneath, fed by the oily drippings, lent a still greater weirdness to the scene.

"Good evenin', gentlemen," said the captain, rising quickly as we entered. "I was settin' here in a sog like, and didn't hear ye. It's a master-night, and we're goin' to have good weather to-morrow. If yaou want to try it agin, ye're welcome." "Sam?

sartin; he's goin'. Him and me's jest like the tigger ten; if yaou haul off the one we ain't good for nothin'. If yaou want to see a faithful friend, jest clap your eyes on Sam Whittlefield. And that ain't all," continued the skipper, looking around and speaking low. " Ye might not think it, for he's master-modest, but Sam's got larnin' that there ain't many in aour taown kin grapple with. Yaou had oughter see his libr'y. A full set of the records of Congress from 1847 to 1861; and he'd have had 'em all, only he jined the navy and couldn't keep 'em up. Then there's a history by Mister Parley, and a hull secretaryful of books of all kinds. Oh, Sam's literary, there ain't no gittin' raound that.

" Yaou might hear him speak of their son John! Wall, he was a chip of the old block. He was as wild a yonker as they make 'em; but Sam never laid the whip on him; he argied with him and eddicated him on a literary principle. When John did anything reckless like, the old lady'd fetch aout a sartin book, called ' The Terrible Sufferin's of Sary Gunley,' an' read him a chapter—like enough ye've heard on't—and I tell ye that tuck the conceit aout of him. She belonged to old Quaker stock, down in Maine, and she kept it up till John was a man grown and she lost her eye-sight. She made a good boy of him; but the poor feller went daown with the rest in the gale of 1875, on the Grand Banks. John had hard luck. The first v'yage he made, the schooner was struck by a sea on the Banks, capsized, and rolled completely under, comin' up the other side, so't the men below dropped out of their bunks on to the ceilin' and then went back ag'in as she righted. The hatches were battened daown and they found John lashed to the wheel, half drowned. The next trip all hands foundered. They reckoned she went daown at the anchorage.

"Have some beans, won't ye?" asked the skipper, abruptly, as if he had been deluded by some trick into a gloomy frame of mind and was determined to shake it off then and there. "Them is the real New England beans," he continued, taking a black bean-pot with a wooden spoon from the ashes. "There's the bone and sinner of New England's sons right here. I'm master fond of 'em; never sails without a pot or so. Every time I see a pot it makes me think of old Joe Muggridge, a deacon of aour taown. He beat me once years ago in 'lection for hog-reeve; but I don't bear no ill feelin'. He was deacon of the First Baptist, and captain of one of the biggest coasters in aour parts, and that fond of beans that folks believed he'd a' died if he couldn't have had 'em. Well, it so happened one fall that there came on a powerful gale on the Georges, and a power of hands was lost. A good many bots got carried away from the schooners, and one dory with two men from Boothbay was picked up by one of these ocean steamers baound in for New York, and that's the way the yarn got aout. They'd been without food and water for three days, and were abaout givin' up; but the steamer folks tuck 'em in, and steamed for port.

"The next mornin' it was blowin' fresh and lively, and the lookaout sighted a schooner lyin' to a couple of miles to the lew'ard, reefed daown close, and a flag flyin' union daown— signal of distress. Thinkin' they were sinkin, the captain of the steamer put towards her, and rounded to half a mile off, and called for volunteers to git aout the bot. Half a dozen brave fellers sprang to the davits, and among 'em aour Boothbay boys. They'd been in a fix, ye see, and was eager to help the rest of sufferin' humanity. She was rollin' so that it tuck 'em nigh an hour to git the bot over, and then two men fell overboard; but finally they got off towards the schooner, all hands givin 'em three cheers.

"It was a hard pull and a nasty sea, but they kept at it, and in half an hour was within hailin' distance. Then the third officer of the steamer stood up and sung aout, 'Schooner ahoy!' 'Ay, ay!' says a man in the schooner's fore-riggin', and the men see naow that she was ridin' like a duck and as dry as a sojer. 'Are ye in distress?' sung aout the officer. 'Yas,' came from the man in the riggin.' 'Founderin'?' shouted the officer agin. 'No,' sung aout old man Muggridge, for it was him' 'next thing to it We're aout o' beans. Kin ye spare a pot?'

"Wall," continued Captain Sol, reddening with the roar of laughter that accompanied the recollection, "it ain't for me, bein' a perfessor of religion, to let on what the men in the bot said, but it had a master effect on the deacon, for afore them rescuers got back to the steamer he'd shook aout his reef and was haulin' to the east'ard.

"Wall," said the old skipper, banking the fire with a shovelful of sand, as his visitors rose to go, "to-morrow, then, at early flood, sharp."

The early flood was that dismal time when the phantom mists of night still cling to the earth, and low-lying clouds of fog cover the river, only to be dispersed by the coming day Cold and cheerless as it was, it found us again launching the whaleboat, and when the sails were trimmed aft and pipes lighted, we rushed into the fog and headed down the river to meet the rising sun.

The mist was so dense that only the glimmer of Captain Sol's pipe could be seen for'ard, appearing like an intermittent eye gleaming through the fog that settled upon our oil-skins in crystal drops and ran in tiny rivulets down the creases into the boat. For a mile we scudded along before the west wind through the gloom, and then a wondrous change commenced. Soft gleams of light shot from the horizon upward, the dark-blue heavens assumed a lighter tint, the pencilled rays growing broader and fusing together, producing a strange and rapidly-spreading nebulous light. The cloud of low-lying mist now became a brassy hue, seemingly heated to ignition, and then from its very substance appeared to rise a fiery, glowing mass that flooded the river with a golden radiance.

"It's a master-sight," quoth Captain Sol between the puffs, as the change went on and the fog began to break before the rising sun. "I ain't no likin' for fogs. Ye see—" But here the skipper stopped, as a peculiar sound and then another, the puffing of the white whale, was heard.

The boat was hauled on the wind, the mast unshipped, and, harpoon in hand, Captain Sol stood braced for the affray. The ripple seemed alive with the ghostly creatures, their white forms darting here and there, while the puffing came fast and furious.

"Stand by to git aft!" whispered the harpooner, and that moment, instead of a white head, the entire body of a Beluga rose in front of the boat, clearing the water in a graceful leap. Quick as thought the skipper hurled his weapon. It struck with a sounding thud, a wing shot, and the great creature fell heavily, impaled in mid-air, to rush away, bearing boat and white-whalers far down the river toward the sea.

INDEX.

INDEX.

ACTINIA, 26, 30, 32.
—— luminosity of, 196.
—— parasitic, 31.
Acrohordus, 121.
Adamsea, 30.
—— on crab, 30.
Agassiz, 24, 56.
Ajuga, 150.
Albatross, 58.
—— pouch of, 58.
Alligator, 54.
Alcyonarian, 47.
—— luminosity of, 47.
Ammocoetes, 6.
Amphicoelias, 107.
Anabas, 36.
—— climbing habits of, 36.
Anolis, 149.
—— mimicry of, 149.
Angler, 25, 210.
Ant, 52.
Ant-eater, 54.
Ancistrodon, 130.
Anableps, 39.
Anemones, 30, 31, 32, 196.
Apeltes, 10.
—— nest of, 10.
Arius, 56.
Argonauta, 59, 163.
Arcturus, 56.
—— care of young, 56.
Argyropelecus, 76.
—— luminosity of, 76.

Architeuthis, 166.
Astrophyton, 168.
Ascidians, 192.
—— luminosity of, 192.
Astrea, 83.
Aspredo, 21, 55.
—— egg of, 55.
Asterias, 26, 46.
—— inhabitants of, 26, 46.
Asteronyx, 48.
—— luminosity of, 48.
Attus volans, 201.
—— flight of, 201.
Atlantic Right Whale, 220.
Atwood's shark, 185.
Auk, 155, 156.
—— extinction of, 155, 156.
Aurelia, 1, 45.
—— luminosity of, 45.

BANDED SUN-FISH, 2.
—— nest of, 2.
Batchelder, J. M., 6.
Batrachian, 9.
Banks, Sir Joseph, 79.
Balloon Fish, 118.
Baird, S. F., Prof., 168.
Barge, 183.
Bagdad, 186.
—— sharks near, 186.
Bead Snake, 131.
Beroidœ, 43.
Bennett, 193.

Beluga, 220.
Belt, 186.
────── on mimicry, 186.
Bibra, 193.
────── on Pyrosoma, 193.
Bison, 161.
────── extinction of, 161.
Blenny, 33.
────── amphibious habits of, 33.
Black snake, 131.
Block Island, 189.
────── large sharks at, 189.
Boleophthalmus, 57.
────── on dry land, 37.
Borneo, 201.
Bryozoon, 145.
Butterfly, 147.
────── mimicry of, 147.
Bulimus, 59.
────── nest of, 59.

CALLICHTHYS, 20, 38.
────── on dry land, 38.
Cancer fulgens, 79.
────── luminosity of, 79.
Carcharias, 186, 191.
────── Fossil, 191.
Cat-fish, 55, 56.
────── eggs of, 56.
────── electric, 118.
────── amphibious, 38.
Campanularia, 15.
────── on nest of Antennarius, 15.
Cantor, 37.
Caudisona, 127.
Camarasaurus, 107.
Caryophillia, 195.
────── luminosity of, 195.
Centipede, 50.
────── care of young, 50.
Cephalopoda, 163.
Ceradotus, 21.

Ceradotus, nest of, 21.
Ceylon, 36.
Chameleon, 149.
────── mimicry of, 149.
Chameleon shrimp, 143.
──────mimicry of, 143.
Chiasmodus, 212.
────── phosphorescence of, 212.
────── method of swallowing, 212.
Chaetodon, 39.
────── long-nosed, 39.
────── shooting drops of water, 39.
Challenger, 35, 56.
Chromataphores, 142.
Chauliodus, 76, 118.
────── luminosity of, 76.
Chlamydoselachus, 103.
Chaetopteridae, 82.
China, 171, 190.
Chrysopa, 209.
────── eggs of, 209.
Chama, 30.
────── boarder in, 30.
Chase, N. D., 92.
────── on sea serpent, 93.
Cleodora, 197.
────── luminosity of, 197.
Clytia, 43.
Clidastes, 106.
Cobra, 128.
Cope, Prof., 103–106.
────── on ancient sea serpents, 105.
────── cretaceous sea serpents, 104.
Crab, 143.
────── mimicry of, 143–145.
Comb-bearers, 43.
Copperhead, 129.
Cowry, 168.
Costa, 11.
Conchs, 172.
Couch, 190.
────── on dog fish, 190.

Crabs, 76, 79, 143.
Craw-fish, 84.
Cricket, 51.
———— nest of, 51.
Crocodile, 54.
Crotalus, 123.
Cuttle fish bone, 163.
Cyclops, 80.
———— luminosity of, 80.
Cyanea, 41.
———— giant, 41.
Cypreas, 168.

DACE, 2.
———— nest of, 2.
Dana, 31.
Daldorf, 36.
———— on climbing perch, 36.
Darwin, 44.
Devil fish, 163.
Dinornis, 160.
———— extinction of, 160.
Dipnoi, 38.
———— burrow of, 38.
Dianea, 44.
———— luminosity of, 44.
Dog fish, 190.
———— a vast number of, 190.
Dodo, 157.
Doras, 38.
———— nest of, 38.
Dorynchus, 80.
———— luminosity of, 80.
Drummond, 43.
Dysmorphosa, 43.
———— luminosity of, 43.

ECHINUS, 30, 54, 84.
Edoux, Prof., 79.
Eel, 32.
———— amphibious habits of, 32.
Elater, 119.

Elater, electric, 119.
Elasmosaurus, 105.
Electric fishes, 113.
———— insects, 119.
Eocene snakes, 137.
Eucope, 43.
Eupomotis, 2.
———— nest of, 2.
Eurypharynx, 211.
———— mouth of, 211.
Exocetus, 198.
———— flight of, 198.
———— blown aboard ships, 198.

FABLES, 134.
Faber, 156.
Fiji, 186.
———— fresh water snakes at, 186.
Fiskenaes, 188.
———— shark fishers at, 188.
Fierasfer, 11, 24.
———— strange habits of, 11, 24.
Flounder, 143.
———— mimicry of, 143.
Flying tree toad, 201.
Florida, 22, 51.
Flying squirrel, 201.
———— lizard, 201.
———— fox, 201.
———— fish, 198.
———— gurnard, 198.
———— gurnard striking sailor, 199.

GAR, 200.
———— striking natives, 200.
———— pike, 33.
———— breathing air, 33.
Garman, Prof., 103.
Gaimard, 24.
Garter-snake, 131, 132.
———— young, in mouth of, 132.
Garden Key, 172.

Geryon, 80.
—— luminosity of, 80.
Gill, Dr., 16, 37.
Giant Crab, 171.
Glaucus, 16.
—— mimicry of, 16.
Gopher, 86.
Gourami, 16.
—— nest of, 16.
Goby, 16, 32, 35.
—— amphibious habits of, 32.
Goode, G. Brown, 71.
—— on sword fish, 71.
—— on snakes swallowing their young, 134.
Goat, 151.
—— intelligence of, 87.
—— games of, 88.
Grouse, 138.
—— mimicry of, 138.
Greef, Dr., 24.
Gryllotalpa, 51.
—— nest of, 51.
Green snake, 135.
Grass fish, 139.
—— mimicry of, 139.
Grunt, 175.
Guadaloupe, 54.
Gurnard, 14, 199.
Gymnotus, 113.
—— electric shocks from, 113.

Haas, 160.
Hassar, 21.
—— amphibious habits of, 21.
Harvey, Rev. Dr., 165.
—— on giant squid, 165.
Hermit Crab, 86.
—— as a pet, 87.
—— in a tobacco pipe, 87.
Hippocampus, 140.
—— mimicry of, 140.

Histiophorus, 69.
—— large fin of, 69.
Hippopotamus, 54.
—— method of carrying young, 54.
Holothuria, 23.
—— mimicry of, 23.
—— inhabitants of, 23.
Hog-nosed snake, 136.
Horned toad, 129.
—— mimicry of, 129.
Holbrook, Dr., 124.
Holder, Dr. J. B., 163.
Hog fish, 135.
Hummingbirds' nest, 139.
—— mimicry of, 130.
Humboldt, 45.
Human mimics, 138.

Ianthina, 15, 145.
—— mimicry of, 145.
—— ink of, 145.
Idotea, 81, 144.
Idylia, 43.
—— phosphorescence of, 43.
Iguana, 149.
Infusoria, 45.
—— luminosity of, 45.
Indians, 161.
—— extinction of,
Ink bearers, 163.
Ink bag, 169.

Jara, 146.
Japan, 171.
—— giant crab of, 171.
Jou Jou, 190.

Kallima, 147.
—— mimicry of, 147.
Key West, 172.

Labyrinthici, 37.
—— habits of, 37.

Lamp fish, 75.
Lamprey, 5.
―――― nest of, 5.
―――― method of building, 5.
Laughing gull, 14.
Latreille, 31.
Labrador duck, 157.
―――― extinction of, 157.
Lasso cell, 26.
Leaf insect, 146.
Leaf cutting ants, 52.
Leydig, 76.
Lemur, 201.
Lizard, 201.
―――― flying, 201.
Liodon, 106.
Locust, 147.
―――― mimicry of, 147.
Lophius, 25, 211.
―――― fishing rods of, 211.
―――― mimicry of, 25.
―――― catching birds, 211.
―――― small fishes in gills of, 25.
Lung fish, 38.
―――― dry burrow of, 38.
Lump fish, 21.
―――― young of, 21.
Lyell, Sir Charles, 91.
―――― on sea serpent, 91.

MALEO, 58.
―――― nest of, 58.
Mammoth, 152.
―――― extinction of, 152.
―――― remains found, 152.
Mastodon, 153.
Mauritius, 157.
―――― birds of, 157.
Maori, 160.
Mantell, 160.
Malapterus, 118.
―――― electric properties of, 118.

Maeandrina, 83.
Marsh, Prof., 106.
Marsupium, 56.
Malthea, 211.
Macrocheira, 171.
Mangrove, 174.
Maneater, 135.
Marston, John, 97.
―――― on sea serpents, 97.
Megapodius, 9, 58, 150.
Melia, 31.
Melicerta, 43.
Medusae, 41.
Metridea, 81.
―――― luminous properties of, 81.
Melanocetus, 212.
Milk snake, 120.
―――― fables about, 120.
Michell, 188.
Mimicry of mammals, 139.
―――― of reptiles, 139, 149.
―――― of fishes, 140.
―――― of crabs, 143, 144.
―――― of ascidians, 145.
―――― of insects, 148.
―――― of plants, 150.
―――― Moa, 160.
Moccasin, 130.
―――― young in mouth of, 130.
Moss insect, 146.
Modiola, 29.
Möbius, Dr., 31.
Mososaurus, 106.
Mole cricket, 51.
―――― intelligence of, 51.
Murenidae, 25.
Mullet, 135.
Myrmelion, 208.
―――― intelligence of, 209.

NAUTILUS, 169.
Nahant, 41, 156.

Naples Aquarium, 24.
—— fierasfer observed at, 24,
Nautilograptus, 144.
—— mimicry of, 144.
New Mexico, 161.
Nestor, 160.
New Zealand, 160.
Newberry, Dr., 163.
Newfoundland, 165.
New Guinea, 190.
Newhall, B. F., 97.
—— on sea serpents, 97.
Noctiluca, 45, 46.
—— luminosity of, 45.
Nototrema, 54.
—— care of young of, 54.
Nordenskiöld, 81, 153.
Nomeus, 28, 140.
—— under physalia, 28.
Notornis, 160.
Notre Dame Bay, 165.
—— giant squid at, 165.
Norbury, Lord, 101.
—— on giant band fish, 101.
Nuttall, 155.
—— on great auk, 155.

Octopus, 166.
—— mimicry of, 166.
—— great size of, 166.
—— strength of, 166.
Olivi, 16.
Onchidium, 207.
Ophiacantha, 47.
—— luminosity of, 47.
Ophidia, 121.
Ophiura, 46.
—— luminosity of, 46.
Ophiocephalus, 37.
Orthoceras titan, 163.
Ostracotheres, 29.
—— parasitic habits of, 29.

Ougel, 32.
—— submarine habits, 32.
Owen, Prof., 160.
Oyster heaps, 152.

Paper Nautilus, 169.
Paradise fish, 18, 28.
—— nest of, 18.
Pavonaria, 48, 49.
—— luminosity of, 49.
Panicum, 17.
Pentacta, 145.
—— mimicry of, 145.
Perch, 2.
—— nest of, 2.
Periophthalmus, 35.
—— amphibious habits of, 35.
Pennatula, 48.
Penguin, 58.
—— pouch of, 58.
Peron, 194.
Pegasus, 200.
—— flight of, 200.
Pelican, 199.
Peripatus, 205.
—— web of, 205.
Pemaquid, 215.
Physalia, 26.
—— author stung by, 27.
Pholas, 196.
—— luminosity of, 196.
Pilot fish, 28–183.
Pinnotheres, 28.
Pipe fish, 140.
—— mimicry, 140.
Pinna, 29.
Pirogue, 187.
—— sunk by rhinodon, 187.
Pliny, 29.
Pleurobrachia, 43.
Pompilus, 200.
—— leaping powers of, 200.

Pollycirrus, 82.
Porgy, 135.
Port Jackson shark, 191.
Portuguese man-of-war, 27.
Porpitw, 15.
Prebloo, 155.
Proven, 188.
Pteranodon, 202.
Ptarmigan, 138.
Pterygotus, 173.
Pyrocistis, 46.

QUAIL, 138.

RATTLESNAKE, 123.
———— young in mouth of, 127.
Remora, 28, 183.
Resolution Island, 161.
Renilla, 47, 108.
———— luminosity of, 108.
Regalicus, 101.
Redi, 109.
Réaumur, 110.
Reduvius, 119.
Rhinichthys, 2.
Rhytina, 153.
———— extinction of, 153.
Rhinodon, 187.
Rhamphorhynchus, 201.
Rhinopoma, 202.
Ribbon fish, 101.
Richardson, Sir John, 12.
Richer, Dr., 31, 113.
Ross, 32.
———— on blenny, 32.
Rumphius, 30.
Ryder, Prof. John, 30.
———— on stickleback, 30.

SALMON, 9.
———— nest of, 9.
Salpa, 16, 145, 193.

Salpa, luminosity of, 145.
———— chains of, 145.
Saury, 75.
Sargasso Sea, 13, 15, 114.
Scopelus, 75.
———— luminosity of, 75.
Scorpion, 53.
———— young of, 53.
Scabra, 24.
Sea anemone, 32.
———— cow, 253.
———— cucumber, 145.
———— horse, 54.
———— serpent, 91.
Selache, 189.
Serpulae, 32.
Semper, 25.
Semotilus, 8.
———— stone heaps of, 8.
Serrasalmo, 20.
———— nest of, 20.
Sepia, 163.
Sea Scorpion, 172.
Seminole, 172.
Stickleback, 13.
———— nest of, 13.
Stichopus, 24.
Sternoptyx, 75.
Surinam toad, 54.
Sword fish, 63.
———— striking boat, 63.
———— ships, 63.

TANGSA, 38.
———— amphibious habits of, 38.
Telfair, 42.
———— on giant jelly fish, 42.
Termitidæ, 52.
Tetrapturus, 70.
Tetraodon, 118.
Thompson, Sir Wyville, 47.
Tinia, 43.

Tima, luminosity of, 43.
Torpedo, 109.
Tortugas, 22.
Toxotes, 38.
—— method of shooting water, 38.
Trap door spider, 148.
—— mimicry of, 148.
Trochus, 85.
Trepang, 23.
Tridacna, 29, 48.
—— sign of, 48.
—— inhabitants of, 49.
Tyrian, 169.

UMBELLULARIA, 48.
—— luminosity of, 48.

VENUS, 49.
Virgularia, 48.
Virgin Islands, 151.

Viti Levi, 186.
—— fresh water sharks at, 186.

WALKING STICK, 146.
Walking leaf, 146.
Wai Levi, 186.
—— sharks at, 186.
White whale, 220.
Woodcock, 57.
—— carry young, 57.
Wortley, Col. S., 30.
Wright, Dr., 187.
Worms, 82.
—— luminosity of, 82.

XIPHIAS, 67.

YUKON RIVER, 20.
—— white whale up, 20.
Yellow tail, 135.